# Research on Vibration Characteristics of Feed System
# 进给系统振动特性研究

LIU Niancong  JIANG Gang
刘念聪  蒋 刚 / 著

四川大学出版社
Sichuan University Press

项目策划：李思莹
责任编辑：蒋　玙
责任校对：余　芳
封面设计：墨创文化
责任印制：王　炜

**图书在版编目（CIP）数据**

进给系统振动特性研究 = Research on Vibration Characteristics of Feed System：英文 / 刘念聪，蒋刚著． — 成都：四川大学出版社，2020.6
ISBN 978-7-5690-3739-5

Ⅰ．①进… Ⅱ．①刘… ②蒋… Ⅲ．①机床－进给－机械振动－研究－英文 Ⅳ．① TG502.14

中国版本图书馆 CIP 数据核字（2020）第 090423 号

| 书名 | 进给系统振动特性研究 |
|---|---|
| | Research on Vibration Characteristics of Feed System |
| 著　者 | 刘念聪　蒋　刚 |
| 出　版 | 四川大学出版社 |
| 地　址 | 成都市一环路南一段 24 号（610065） |
| 发　行 | 四川大学出版社 |
| 书　号 | ISBN 978-7-5690-3739-5 |
| 印前制作 | 四川胜翔数码印务设计有限公司 |
| 印　刷 | 成都市新都华兴印务有限公司 |
| 成品尺寸 | 170mm×240mm |
| 印　张 | 7 |
| 字　数 | 176 千字 |
| 版　次 | 2020 年 6 月第 1 版 |
| 印　次 | 2020 年 6 月第 1 次印刷 |
| 定　价 | 48.00 元 |

◆ 版权所有 ◆ 侵权必究

◆ 读者邮购本书，请与本社发行科联系。
　电话：(028)85408408/(028)85401670/
　(028)86408023　邮政编码：610065
◆ 本社图书如有印装质量问题，请寄回出版社调换。
◆ 网址：http://press.scu.edu.cn

四川大学出版社
微信公众号

# Preface

In the world, the development of the modern industry is inseparable from the advanced manufacturing technology. As a major core pillar of the steady development of industry, it has been highly valued by the major industrial countries. For advanced manufacturing technology, CNC machine tools, which are not only suitable for the aerospace field, but also play an important role in industrial manufacturing industries, are indispensable equipment.

As one of the key subsystems of CNC machine tools, the performance of feed system can directly affect the stability of machine tools to a large extent, the surface quality of workpiece and tool life, especially in high-end manufacturing equipment such as ultra-precision machine tools. Nowadays, the related technologies in this field have attracted the general attention of industrial circles and academia. It has broad development prospects and long-term vitality. This book covers theory, numerical simulation and experiment of the vibration analysis of the feed system of CNC machine tools. It provides not only the necessary basic knowledge for conducting the vibration analysis of the feed system, but also the latest methods and results for its vibration control. The book also provides a useful reference for the analysis and control of mechanical vibration. We sincerely hope that this book will benefit machine tool design engineers and researchers in other related fields.

This book is divided into 8 chapters. Chapter 1 briefly introduces the development history and research status of vibration technology of machine tool feed system. In Chapter 2, the method of calculating the mode of the screw, which is the key part of the ball screw feed system is introduced. The modeling technology based on work-energy theorem, precise finite element simulation technology and modal measurement experiments are introduced. Chapter 3 presents the vibration analysis technology based on rigid flexible coupling technology. Chapter 4 introduces the influence law of the stiffness factors on the vibration of the feed system, mainly including static stiffness and dynamic stiffness. In Chapter 5, the influence of dynamic stiffness on vibration of feed system is studied theoretically and experimentally. Chapter 6 investigates the influence of damping property on chatter of feed system, which provides the necessary knowledge for the further study of the vibration law. The axial non-linear vibration of the feed system is introduced in Chapter 7. The theoretical model and corresponding control strategy are established in Chapter 8. In this book, except Chapter 1, the mechanical models of the feed system are established, and the simulation analysis and verified experiments are carried out. Each chapter introduces the newest research technique and the progress in the related fields in this book.

This book is written by Liu Niancong and Jiang Gang. Chapter 1~2 and Chapter 4~8 are written by Liu Niancong, Chapter 3 is written by Jiang Gang, and illustrations are revised by Yuan Jia, Zou Xing, Yang Chengwen, etc. Professor Xie Jin reviewed the whole book.

This book is supported by the National Natural Science Foundation of China (Grant No. 51575457), the Sichuan Science and Technology Program (Grant No. 2019YFG0358), the Sichuan Science and Technology Program (Grant No. 20ZDZX0013), and the Project of Chengdu Science and Technology Bureau (Grant No. 2015−NY02−00285−NC). This support is gratefully acknowledged. Many thanks are due to the referees for their valuable comments.

Thanks to Professor Xie Jin at Southwest Jiaotong University for his help with research on non-linear vibration of feed system. Also thanks to Yuan Jia, Zou Xing, Yang Chengwen, Wang Jin, and other master students for their hard work in proofreading this book.

The publication of this book has been greatly supported and helped by Sichuan University Press, and the authors express their heartfelt thanks. Part of the book adopts the research results of the authors' team, and especial authors thanks to the researchers who have worked together with the editors and contributed to these research results. There might be mistakes and defects in the book. Authors are willing to accept the shortcomings pointed out and make improvements.

# Contents

**Chapter 1　Introduction**　　( 1 )

1.1　Machine tool vibration　( 1 )

1.2　Influence of feed system vibration　( 2 )

1.3　The current status of feed system vibration　( 4 )

　　1.3.1　The establishment method of analytical model　( 4 )

　　1.3.2　The method of vibration control　( 10 )

References　( 12 )

**Chapter 2　The Vibration Modal Analysis of the Ball Screw about Machine Tools**　( 15 )

2.1　Introduction　( 15 )

2.2　Physical model of feed system　( 16 )

2.3　Mechanical model of ball screw　( 16 )

2.4　Analysis of speed ball screw based on FEM　( 20 )

2.5　Modal experiment of ball screw ……………………… ( 23 )

    2.5.1　Modal experiment scheme ……………………………… ( 23 )

    2.5.2　Analysis of experimental results ……………………… ( 24 )

References ……………………………………………………………… ( 25 )

## Chapter 3　Rigid-flexible Coupling Model and Dynamic Characteristic Analysis of Feed System ………… ( 27 )

3.1　Introduction ……………………………………………………… ( 27 )

3.2　Mathematical modeling ………………………………………… ( 27 )

    3.2.1　Dynamic model of linear guide pair ………………… ( 27 )

    3.2.2　Dynamic model of ball screw ………………………… ( 28 )

3.3　Establishment of rigid-flexible coupling model ………… ( 29 )

    3.3.1　Establishment of the flexible body model …………… ( 29 )

    3.3.2　Rigid body model establishment and simulation parameter setting ……………………………………………………… ( 30 )

3.4　Analysis of simulation results ………………………………… ( 31 )

    3.4.1　Correlation analysis ……………………………………… ( 31 )

    3.4.2　The influence of damping coefficient and stiffness on vibration ……………………………………………………… ( 32 )

References ……………………………………………………………… ( 33 )

## Chapter 4　Static Stiffness Analysis of Feed System ……… ( 35 )

4.1　Introduction ……………………………………………………… ( 35 )

4.2　Mathematical model of axial static stiffness …………… ( 36 )

    4.2.1　A universal model of axial static stiffness of ball screw pair ……………………………………………………… ( 36 )

4.2.2　Axial static stiffness of nut ................................. ( 37 )

4.2.3　Axial static stiffness of angular contact ball bearings ...... ( 39 )

4.2.4　Axial static stiffness caused by torsional deformation of the screw ................................................. ( 39 )

4.3　Validation of axial static stiffness model ..................... ( 40 )

4.4　Influence of various factors on axial static stiffness ...... ( 42 )

4.4.1　Axial load ................................................. ( 42 )

4.4.2　Number of balls ........................................... ( 43 )

4.4.3　Contact angle ............................................. ( 44 )

4.4.4　Helix angle of the lead screw ............................. ( 46 )

4.4.5　Displacement of the workbench ............................. ( 46 )

4.4.6　Rigidity of each component ................................ ( 47 )

References ........................................................ ( 49 )

## Chapter 5　Influence of Dynamic Stiffness on Vibration of Feed System ............................................................. ( 51 )

5.1　Introduction ................................................. ( 51 )

5.2　Dynamic model considering coupling stiffness ................. ( 52 )

5.3　Experimental verification of dynamic model ................... ( 58 )

5.4　Vibration characteristic analysis ............................ ( 61 )

References ........................................................ ( 66 )

## Chapter 6　Influence of Damping Property on Chatter of Feed System ............................................................. ( 68 )

6.1　Introduction ................................................. ( 68 )

6.2　Dynamic model based on the Lagrange equation ................. ( 69 )

6.3　Simulation analysis ……………………………………… ( 72 )

 6.3.1　Simulation parameter setting ……………………………… ( 72 )

 6.3.2　Simulation result analysis ……………………………… ( 73 )

References ……………………………………………………… ( 79 )

**Chapter 7　Axial Non-linear Vibration of Feed System** ………… ( 81 )

7.1　Introduction ……………………………………………… ( 81 )

7.2　Theoretical modeling ……………………………………… ( 82 )

7.3　Numerical simulation ……………………………………… ( 85 )

 7.3.1　Ball-screw length $L$ ……………………………………… ( 86 )

 7.3.2　Exciting force $F$ ………………………………………… ( 86 )

 7.3.3　Excitation frequency $\omega$ ……………………………… ( 87 )

 7.3.4　Damping ratio $c$ ………………………………………… ( 88 )

7.4　Experimental verification ………………………………… ( 89 )

References ……………………………………………………… ( 90 )

**Chapter 8　Adaptive Control of Feed System Based on LuGre Model**
 ……………………………………………………………… ( 91 )

8.1　Introduction ……………………………………………… ( 91 )

8.2　Dynamic modeling of the feed system ………………… ( 92 )

8.3　Designing of the friction compensation controller …… ( 94 )

8.4　Simulating of adaptive friction compensation ………… ( 96 )

References ……………………………………………………… (103)

# Chapter 1    Introduction

## 1.1    Machine tool vibration

The vibration of the machine tool affects its stability, the quality of the workpiece, and the tool's life. According to the kinds of excitation, the vibration can be divided into the external excitation vibration (the forced vibration) and the self-excited vibration (the chatter). The external excitation vibration mainly depends on the form of the machine tool transmission, the assembly accuracy, the dynamic characteristics of the motor, the foundation, the outside environment and so on. And the external excitation is influenced by the cutting parameters, the material characteristics of the workpiece, the parameter and the performance of the cutting tool and so on. In engineering, most people think that the vibration in the cutting process has a bad effect on the quality of the workpiece and the tool's life. In the cutting process, the cutting parameters are always forced to change, for example, the cutting depth is reduced, the feed rate is altered and so on, to decrease the bad impact of the vibration. Then it will lead to higher manufacturing costs[1].

In the 1940s, professor R. N. Arnold at the University of Edinburgh firstly studied the cutting vibration both experimentally and theoretically. He observed the vibration phenomenon of the lathe tool in the velocity direction through the experiments. And he thought that it was the main reason for the chattering that the cutting force continually decreased with the increasing velocity. Then he analyzed the impact of the vibration on the regeneration of the built-up edge, tool wear, the vibration mark of the workpiece surface, tool's life and so on. In the 1950s, the study mainly focused on the analysis of various cutting vibration phenomena. What's more, many kinds of assumptions and theoretical models were put forward to explain the reason and the influence factors of the chattering. And the research focused on three issues: the cutting process modeling, the structure of the machine tool modeling, and the stability analyzing. Though the study of the vibration in the machine tool has lasted for half a century, the reason and the mechanism of the chattering can not be explained well until now. And there are still a lot of vibration phenomena that can not be explained.

## 1.2 Influence of feed system vibration

The feed system is characterized by high rigidity and low sensitivity to the outside disturbance and it has always been the key component of the Numerical Control machine tool (NC machine tool). And its dynamic characteristics affect and restrain the machining accuracy and efficiency of the NC machine tool significantly. Nowadays, the mechanical transmission system of the NC machine tool generally consists of the motor, the ball screw, the nut, and the workbench. Bearing and the guide rail are the main supporting part. Consequently, the whole system is a flexible one which is composed of many parts. The dynamic characteristics of the mechanical transmission are influenced both by every component and the connection between the components. Especially, there will be more prove to be vibration during high velocity or high acceleration feeds. With the increasing demand of the velocity and accuracy of the NC manufacturing, the influence of the feed system dynamic characteristics on the accuracy of the NC manufacturing becomes increasingly prominent.

The vibration of the feed system has a significant impact on the quality of the workpiece surface and the tool's life. During metal cutting process, there will always be strong vibration between the workpieces and the cutting tools, so this vibration is often considered as a harmful phenomenon which can destroy the normal process. When the cutting occurs, the quality of the workpiece surface gets serious deterioration and the strip or scaly ripple will appear on the surface of the workpiece. The ripple strongly affects the finish of the machined surface and makes the value of surface roughness increase. At this time, the rise of productivity is restrained by having to decrease the cutting parameters. Furthermore, the vibration will also increase the tool wear. If the vibration gets intensive, the edge of the cutting tool will be chipped. Then it will make the process stop. Lots of experiments and practice manifest that the vibration of the feed system will impact on the surface quality of the workpiece and tool's life directly. But the law will be different with different cutting conditions[2-3].

An interesting phenomenon which is worth noting is that a new vibration-assisted cutting technology can decrease the surface roughness or increase the tool's life by using specially conditional vibration[4-5]. But is still unknown that in which conditions vibration can be beneficial.

The vibration of the feed system has significant impacts on the stability of the NC machine tool[6]. The quality of the machining part in the NC machine tool is directly decided by the stability of the accuracy of the NC machine tool, and lots of research institutes and firms propose different kinds of methods to increase the positioning accuracy of the machine tool. With the increase of the velocity and acceleration and the increasing demand of the machine tool accuracy, the influence of the mechanical system flexibility (finite rigidity of transmissions) and the load inertial on the feed system is more outstanding. Then the fixed combination part of the machine tool is the important basic structure in the machine tool and the stability of the combination part has a fundamental effect on the rigidity and dynamic stability of the whole machine. Machine tools are composed of various parts according to certain requirements, so there are many combined parts. The combined parts in the machine tool can be divided into two types: the moving combination and the static (fixed) combination. The moving combination is just like the combination of the skateboard and the machine bed rail, shaft and bearing and so on. The static combination is just like the combination of the upright and the machine tool bed, the box, and the machine tool bed. Because the contact pressure at the connecting surface is always limited in a range, it can not be infinite, furthermore, the geometric errors and the micro irregularities always exist at the connecting surface and there are lubrication films at some connecting surface. Under the outside force, the motion of the parts in the machine has periodicity, which is the characteristic of the elastic-deformation process. Therefore there will be vibration and the periodic inertial force, and their effects appear as the form of additional loading. At the same time, they will also cause deformation. When the outside forces cause the deformation stop, the parts try to get back to their original positions. However, because of the energy collecting, the parts get to the oppositional directions, and then go back towards the original directions and so on. In this way, the vibration whose amplitude is much smaller than the dimension of the parts takes place. Because of the vibration, there will be chattering and impact between the parts. What's more, the heat quantity between the bearing and shaft, the wear of the friction parts are also increasing. The transmission power and the efficiency decrease because of consuming energy in the vibration. The intensive vibration may cause stress beyond the limited stress of the parts and make the parts destroyed[7].

## 1.3 The current status of feed system vibration

### 1.3.1 The establishment method of analytical model

The establishment methods of the dynamic model in the machine tool feed system mainly include the analytical method, finite-element method, matrix method and so on. Many researchers have done a lot of fruitful work.

(1) Dynamic modeling by the centralized mass parameter method.

Aiming at the periodic variation of the cutting force in the milling and the dynamic characteristics of multi-degree of freedom structure, Engine and Altintas[8] proposed a vibration model of the dynamic milling system with multi-degree-of-freedom structures. Considering the rigidity and damping of the processing system and using the Fourier series expansion of the milling force coefficients, they proposed the dynamic model and general mathematical expression to forecast the stability of the milling. Choi et al.[9], by analyzing the vibration characteristics of the ball screw feed system at the plane motion, proposed a kind of centralized parameter model with six degrees of freedom and inferred the kinematical equation of this model. And they analyzed the natural frequency of the system and the transient response caused by the input of the driving motor velocity controlling, then they took model analysis and measured the working vibration of the ball screw in the laboratory. Finally, both the theoretical analysis and the measurement results in the experiment appear to have good consistency. Yang et al.[10] used the D'Alembert's principle to model the mechanical system, and carried out theoretical analysis and experimental research on the mechanical dynamics characteristics and influencing factors of the linear motor table. Although the centralized mass parameter method decreases the number of the degree of the freedom of the model and decreases the order of the transfer function efficiently, some important factors which can affect the performance of the system, such as the flexibility of the system structure, the rigidity of the connecting parts and so on, will be restrained by the more simplified research method and the model ways. At the same time, the slender rod characteristics of the ball screw will also be ignored.

(2) Dynamic analysis by the finite-element method.

Fu[11] has proposed a finite-element modeling method. He has simplified the ball screw, the rolling guide, and bearing as the spring-mass model. What's more, he inferred the formula of calculating the rigidity of each connecting surface based on the Hertz contacting formula and got the value of the contacting rigidity in a status. Considering the influence of the torsional rigidity of the screw and the coupling, the dynamic model of the whole feed system has been built by using finite-element software. Then the first five order modal vibration modes and natural frequency of the loading platform have been analyzed in a simulating way. The result of the finite-element simulating has been verified by the modal experiments. Zaeh and Oertli[12] regarded the screw as the constant section beam to research its dynamic characteristics of the feed system by using the finite-element method. They used the finite-element method to analyze the dynamic characteristics of the ball screw feed system of the NC plane milling and boring machine, and discussed the modeling method of the contacting surface in finite-element models. According to the analysis of the modal and harmonic response, the natural frequency and the vibration characteristics of the machine tool feed system have been obtained. So it provided a reliable basis for optimizing the structure of the feed system and improving the dynamic characteristics. And the results of the finite-element analysis were verified through the experiment[13]. The dynamic model built by the infinite-element method is more complicated and the number of degrees of freedom in the model is too much. What's more, the model is also a little larger and it is not convenient for the servo system to integrate controlling. At the same time, the torsion rigidity of the coupling and the ball screw is often ignored when feed system is modeling. In order to facilitate the integration of the control system and the mechanical transmission, the ball screw feed system is often simplified as a simple mechanical model. But there are obvious shortages in predicting exactly the characteristics of the structural dynamics and accurately solving the response.

(3) Modeling method by elastic mechanics.

This method mainly includes Hertz contacting and displacement. The former is used to analyze the influence of the connecting parts parameters on the feed system vibration. For example, Sun modeled the linear rolling guide and solved the contacting rigidity problem based on the Hertz contacting principle. He inferred the analytical expression of the contacting rigidity of the single ball-groove in the whole guide rail system and systematically analyzed the deformation relationship between the slider and the guide rail. Finally, he finished the vertical contacting modeling of the whole guide rail. Using this model, the influence of the external load and preload on the rigidity of the guide rail has been analyzed[14]. But this method is restrained by the accuracy of the Hertz contacting theory and there is a big error in the practical. Therefore this modeling method needs to be improved.

The displacement method is generally used to analyze the non-linear vibration problem of the feed system. For example, Wu used the elastic mechanics displacement method and Galerkin method to build the mathematical model of the feed system ball screw. And he also used the Poincare-Lindstedt singular perturbation method to obtain the approximate analytical solution of the quadratic free vibration in the system. The model has been verified by the parameters identification simulating and experimental research. During the movement of the feed system, it is found out that the non-linear vibration appears in the ball screw system under the influence of the axis force, radial force, torque and many loads and there will also be forked and chaos phenomenon. Furthermore, the rigidity of the system will change with the different positions of the workbench and it appears non-linear law. According to the time series of the workbench vibration acceleration and the chaos characteristics shown in the spectrums, the feed system can be determined as a non-linear dynamic system[15]. Yang et al. [16] built the dynamic model of the screw-workbench system based on the assumption of the Timoshenko beam and considering the transverse shearing of the screw shaft. According to the analysis, the non-linear and coupling factors existed between every direction vibration of the screw shaft. Furthermore, the transverse vibration can be expressed by the forced vibration Duffing equation. The forced excitation came from the influence of the bending vibration coupling and its cause is the shearing deformation of the beam under the Timoshenko assumption. Through analysis of the main component of the dynamic characteristics, the dynamic model of the ball screw and the description of its characteristics are verified. Wang et al. [17], through the theoretical analysis and experimental research, focused on the effects of the non-linear spring force and non-linear friction on the NC workbench dynamic characteristics. The NC workbench was influenced by the axis force, lateral force, torque, friction, cutting force and many other loads. The value of the kinds of the rigidity of the ball-screw was closely related to the supporting way of the ball-screw. The value of every kind of rigidity changed with varying of the displacement and moving the direction of the workbench and it appeared soft spring characteristics or hard spring characteristics and other non-linear laws. The changing law of the friction obeyed the Streibeck curve. It pointed out that the effect of the non-linear spring force can be described by using the Duffing equation with damp, and the effect of the non-linear friction can be described by using the Vander Pol equation. And the coupling effect between the non-linear spring force and non-linear friction can be described by using the Lienard equation. Using the displacement method, modeling can be complex and solving gets to be very difficult.

(4) Hybrid modeling.

Considering the limitation of the above method, more and more researchers use hybrid modeling methods to analyze the dynamic characteristics of the system. Chinedum definitely took the lateral dynamics of the screw into the hybrid finite-element model of the ball-screw transmission. The ball screw was modeled by the Timoshenko beam element. Furthermore, the ball, coupling, bearing, and fastener were regarded as pure springs. The rigid member is modeled by using the centralized mass method. Considering the influence of the lateral vibration, Chinedum and Okwudire[18] proposed a screw-nut interface model to predict the coupling effect of the axis, torsion and lateral dynamic of the ball-screw. And the influence of this dynamic coupling effect on the driving position accuracy has been verified by experiment. Dong et al. [19] proposed a kind of hybrid model method and used the power balanced method and Ritz series method to build the vibration model of the axis direction, torsion, and bending of the ball screw feed system. And they deduced the matrixes of the rigidity, mass and damp in the system. Li et al. [20], considering the centralized flexibility and distributed flexibility characteristics, built the electromechanical-rigid flexibility coupling dynamic model of the feed system, and discussed response characteristics of the step signal and sine signal to the feed system. What's more, the influence law of generating the process of the linear feed system dynamic errors and other kinds of factors on dynamic errors has been analyzed. Mi, based on the Hertz contacting theory, calculated the influence of the preload on the axis rigidity of the ball-screw and discussed the influence of the moment of pretention on the stress of the bolting connecting. Furthermore, using the optimization algorithm, the dynamic rigidity of the line guide rail and the damp were simulated and experimentally verified. The finite-element model of the whole machine tool, affected by different connecting ways, was built. What's more, the experimental result was verified and the influence of the preload on the ball-screw and line guide rail was predicted. The results have shown that the preload of the machine tool connecting head has an obvious effect on the dynamic rigidity of the end of the main shaft[21].

Besides, many scholars also researched the influence of the structure parameters of the feed system on the system dynamic characteristics. In 2010, Dadalau et al. thought that, in the ball-screw transmission, the dynamics behaviors were mainly determined by the geometry of the ball screw itself. The performances of the axial and torsion rigidity, the moment of inertia, maximum velocity and acceleration were not only determined by the servomotor, but also the diameter and slope of the screw and the radius of the ball grooves. To analyze these effects, a new method to calculate the axial and torsion rigidity of the ball screw has been proposed. The parameter equation has been analyzed and deduced and the dependence relationship between the rigidity and the geometry parameters of the screw has been analyzed[22]. In 2018, Ohashi et al. [23] analyzed the influence of the linear ball guide rail preloading and the ball retainer on the micro-scale feed driving system based on the line guide rail system built by themselves (Fig. 1-1). Through the establishment of the friction controlling system model, the time constants of each step response are analyzed. The analysis and experiment verified that the influence of the carriage friction characteristics on the quadrant errors and the step response behavior was very significant. Furthermore, the research has shown that the friction characteristics also have some effect on the steady-state vibration and the amplitude was directly proportional to the flexibility of the non-linear spring behaviors.

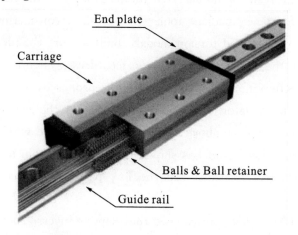

Fig. 1-1  SPS25 eight grooves linear ball guide rail sketch map

## 1.3.2 The method of vibration control

The vibration of the feed system is generally controlled by using matching the parameters of the system structure or the feed controlling strategies. The modern NC machine tool, in the process, is demanded to make full use of the feed system velocity and acceleration, and make sure to have high position precision in high speed. The research of literature[24] has shown that the dynamic errors caused by the feed system, compared to the static geometry errors, were much larger when the NC machine tool was working in high speed and acceleration.

To improve the performance of the servo control system, there is a need to use advanced technology to improve the dynamic characteristics of the feed system, even the whole servo control system. Because the acceleration and deceleration control strategy of the ball screw driving mechanisms feed motion and the theoretical analysis model of the feed system vibration response is not mature enough, the continuity of the movement command curve is only improved by the experience of the operators. To some degree, the above way can decrease the feed vibration generated by the high speed, acceleration feed process.

To decrease the machine tool vibration, Hiramoto et al.[25] researched into the effectiveness of the barycenter driving principle in vibration reduction. According to the results, he analyzed the possibility of the new designing of the higher performance machine tool. Tsai et al.[26] considering the machine tool bed, and vertical and torsion modal, built a more completed model. In order to reduce the vibration of the servo feed drive system, a zero-vibration full-order correction input shaping algorithm is proposed, which is integrated in to the NC interpolator. The vibration restrains the performance of the inputting shaper using the above algorithm was compared with the traditional inputting shaper. The result has shown that this algorithm not only had the notch filtering effect but also had the low pass filter effect in the high-frequency mode.

Fallah and Moetakef[27] researched the adaptive inverse control problem of the internal cylindrical turning chattering. It includes using the active boring bar as the controlled actuator and the IEPE accelerometer as the feedback transducer. They proposed a kind of new adaptive inverse control algorithm and verified the effectiveness during the experiment of internal turning aluminum alloys 6064-T6. Around the peak of the natural resonance of the active boring bar, the amplitude of the chattering has been restrained effectively. At the same time, the periodicity chattering traces in the texture of the workpiece surface were removed. And the roughness of the cutting surface has been improved obviously. Luo et al.[28], against the vibration problem of the NC machine tool feed system caused by the discontinuous movement commands, built the theoretical model of the servomotor torque excitation and feed system response based on the centralized mass dynamics modal of the ball screw feed system. And the relationship between the movement commands acceleration curve power spectrum and the feed system vibration response has been analyzed. Aiming at the vibration excitation problem introduced by the discontinuous start point and the endpoint of the acceleration curve in the acceleration and deceleration control strategy, a kind of cosine acceleration and deceleration control strategy has been proposed. And the above strategy realized the smooth transition of the whole acceleration process. Wang et al.[29], to overcome the difficulty control problem of the ball screw vibration, proposed a method of controlling the lateral vibration of the ball screw using a multiple tuned mass damper. They built the dynamic model on the condition that the nuts and the ends of the bearing are elastic supporting, of ball screw including multiple tuning mass damper. And it obtained the movement with the different nuts and the lateral vibration frequency response function excited by the dynamic working loads. To optimize the maximum and minimum amplitude of the different nuts position screw lateral vibration frequency function, the optimal dynamics parameters of the multi tuning mass dampers were obtained. The research results have shown that the effects and the robustness were more obvious by the screw lateral vibration controlling of the multi tuning mass damper. Gong et al.[30], based on the modal reduction method, built the electromechanical coupling model of the machine tool feed system. And comparing the simulation and the experiment, the effectiveness of the established electromechanical coupling simulating model is verified. Based on the feed system electromechanical coupling model, by comprehensively using the experiment designing, response surface fitting, and multi-objective optimization algorithms, and taking the bearing preload and servo control parameters as variables, the dynamic characteristics of the horizontal machining center feed system have been optimized. The results have shown that the position accuracy of the $X$, $Y$, $Z$ axes and the efficiency of the response have increased slightly. What's more, the axial amplitude of the $X$, $Y$, $Z$ axes have decreased by 70.08%, 63.76%, 79.32%, so the dynamic characteristics have increased obviously.

From the above, to realize the vibration control of machine tools, the dynamic characteristics of machine tools must be deeply analyzed. Even in the active design, it is necessary to understand the influence of various factors on the dynamic characteristics of the feed system, such as the influence factors on the vibration mode and amplitude. In addition, due to the complex coupling and restriction relationship between various factors, reasonable structural design and stiffness distribution should be adopted in the design stage to achieve the purpose of matching with various factors.

# References

[1] Li H Z, Spur G. Review and prospect of the machine tool vibration research [J]. Journal of Tongji University (Natural Science), 1994, 22 (3): 378-384.

[2] Ulas H B, Ozkan M T, Malkoc Y. Vibration prediction in drilling processes with HSS and carbide drill bit by means of artificial neural networks [J]. Neural Computing & Applications, 2019, 31 (9): 5547-5562.

[3] Liu N C, Zheng C L, Xiang D Y, et al. Effect of cutting parameters on tool wear under minimum quantity cooling lubrication (MQCL) conditions [J]. The International Journal of Advanced Manufacturing Technology, 2019, 105 (1-4): 515-529.

[4] Brehl D E, Dow T A. Review of vibration-assisted machining [J]. Precision Engineering, 2008, 32 (3): 153-172.

[5] Xu J Y, Li C, Chen M, et al. A comparison between vibration assisted and conventional drilling of CFRP/Ti6Al4V stacks [J]. Materials and Manufacturing Processes, 2019, 34 (10): 1182-1193.

[6] Peng Y L, Li B, Mao X Y, et al. Characterization and suppression of cutting

vibration under the coupling effect of varied cutting excitations and position-dependent dynamics [J]. Journal of Sound and Vibration, 2019 (463): 1-17.

[7] Guo Y, Lin B, Wang W Q. Optimization of variable helix cutter for improving chatter stability [J]. The International Journal of Advanced Manufacturing Technology, 2019, 104 (5-8): 2553-2565.

[8] Engine S, Altintas Y. Mechanics and dynamics of general milling cutter part: Helical end mills [J]. International Journal of Machine Tools and Manufacture, 2001 (41): 2292-2231.

[9] Choi Y H, Cha S M, Hong J H. A study on the vibration analysis of a ball screw feed drive system [J]. Advances in Materials Manufacturing Science and Technology, 2004 (471-472): 149-154.

[10] Yang X J, Zhao W H, Liu H, et al. Dynamic characteristics of mechanical system in linear motor feed system [J]. Journal of Xi'an Jiaotong University, 2013, 47 (4): 44-50.

[11] Fu Z B, Wang T Y, Zhang L, et al. Dynamic modeling and vibration analysis of ball screw feed driving systems [J]. Advances in Materials Manufacturing Science and Technology, 2019, 38 (4): 56-63.

[12] Zaeh M F, Oertli T. Finite element modelling of ball screw feed drive systems [J]. CIRP Annals-Manufacturing Technology, 2004, 53 (2): 289-292.

[13] An Q Y, Feng P F, Yu D W. Analysis of dynamic characteristic of ball screw feed system based on FEM [J]. Manufacturing Technology & Machine Tool, 2005 (10): 85-88.

[14] Sun W, Kong X X, Wang B, et al. Contact modeling and analytical solution of contact stiffness by Hertz theory for the linear rolling guide system [J]. Engineering Mechanics, 2013, 30 (7): 230-234.

[15] Wu Q, Rui Z Y, Yang J J. Bifurcation and chaos characteristics for ball-screw system considering non-linear elastic force [J]. Journal of Xi'an Jiaotong University, 2012, 46 (1): 70-75.

[16] Yang Y, Zhang W M, Zhao H P. Dynamic characteristics of ball screw system [J]. Journal of Vibration, Measurement & Diagnosis, 2013, 33 (4): 664-669.

[17] Wang L H, Du R S, Wu B, et al. Nonlinear dynamic characteristics of NC table [J]. China Mechanical Engineering, 2009, 20 (13): 1513-1519.

[18] Chinedum E, Okwudire Y A. Hybrid modeling of ball screw drives with coupled axial, torsional, and lateral dynamics [J]. Journal of Mechanical Science and Technology, 2009, 131 (7): 10021-10029.

[19] Dong L, Tang W C, Liu L. Hybrid modeling and time-varying analysis of

vibration for a ball screw drive [J]. Journal of Vibration and Shock, 2013, 32 (20): 196-202.

[20] Li J, Xie F G, Liu X J, et al. Dynamic modeling of a linear feed axis considering the characteristics of the electro-mechanical and rigid-flexible coupling [J]. Chinese Journal of Mechanical Engineering, 2017, 53 (17): 60-69.

[21] Mi L, Yin G F, Sun M N, et al. Effects of preloads on joints on dynamic stiffness of a whole machine tool structure [J]. Journal of Mechanical Science and Technology, 2012, 26 (2): 495-508.

[22] Dadalau A, Mottahedi M, Groh K. Parametri cmodeling of ball screw spindles [J]. Production Engineering, 2010, 4 (6): 625-631.

[23] Ohashi T, Shibata H, Futami S. Influence of linear ball guide preloads and retainers on the microscopic motions of a feed-drive system [J]. Journal of Advanced Mechanical Design Systems and Manufacturing, 2018, 12 (5): 1-10.

[24] Feng B, Mei X S, Yang J, et al. Adaptive compensation of friction error for numerical control machine tool [J]. Journal of Xi'an Jiaotong University, 2013, 47 (9): 65-69.

[25] Hiramoto K, Hansel A, Ding S. A study on the drive at center of gravity (DCG) feed principle and its application for development of high performance machine tool systems [J]. Cirp Annals-Manufacturing Technology, 2005, 54 (1): 333-336.

[26] Tsai M S, Huang Y C, Lin M T. Integration of input shaping technique with interpolation for vibration suppression of servo-feed drive system [J]. Journal of the Chinese Institute of Engineers, 2017, 40 (4): 284-295.

[27] Fallah M, Moetakef I B. Adaptive inverse control of chatter vibrations in internal turning operations [J]. Mechanical Systems and Signal Processing, 2019, 129 (15): 91-111.

[28] Luo L, Zhang W M, Jürgen F. Excitation response characteristics of ball screw feed drive system and cosine-like jerk motion profiles [J]. Journal of Vibration, Measurement & Diagnosis, 2019, 39 (1): 160-167.

[29] Wang M, Li F J, Zan T, et al. Bending vibration control of a ball screw via multi-tuned mass dampers [J]. Journal of Vibration and Shock, 2013, 47 (11): 65-69.

[30] Gong Z C, Niu W T, Li J J, et al. Dynamic characteristics simulation and performance optimization of the feed system based on the electromechanical coupling [J]. Journal of Machine Design, 2018, 35 (9): 8-16.

# Chapter 2  The Vibration Modal Analysis of the Ball Screw about Machine Tools

## 2.1  Introduction

With the increasing demands of the high-speed, and high-accuracy of the ball screw, the vibration modal analysis is applied in the fault diagnosis and prediction, optimization of structural dynamic characteristics, prevention of resonance and self-excitation vibration increasingly, and has become one of the focuses at home and abroad[1-3].

As researchers at home and abroad analyze the dynamic characteristics of ball screw, the theoretical methods used in the study mainly include: ①Based on Timoshenko beam, Euler-Bernoulli beam, the dynamic model was constructed. For example, the dynamic model of machine feed system was built based on Timoshenko beam, and the coupling vibration relations of the screw in different direction were analyzed[4-5]. Accused of neglecting the impact stiffness loss of screw thread, analytical error is bigger. ②The finite element method (FEM). The FEM is divided into two categories based on whether the characteristics of the thread are ignored. The effects of dynamic characteristics for screw thread features were adequately considered[6-7]. The simplified model of ball screw was directly adopted. For small pitch and deep groove ball screws, if the effect of thread characteristics on damping is ignored, the elastic coefficient will tend to have a larger error[8-10].

## 2.2 Physical model of feed system

Fig. 2-1 shows the structural principle of ball screw feed system of CNC machine tool. Feed system composition mainly includes screws, tables, sliders, guides, bearings, etc. It is to drive the slide and workbench to move along the guide in a straight line through the ball screw pair, so as to realize the accurate positioning and feeding of the table in the $x$-axis.

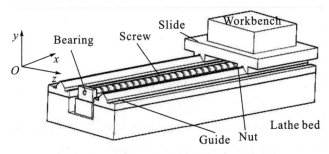

Fig. 2-1  Structure of machine tool feed system

## 2.3 Mechanical model of ball screw

Ball screw structure, with thrust bearings on one side and radial ball bearings on the other side, is considered to be a static and uncertain structure with one end hinged and the other end rigidly connected. For the ball screw, the heterogeneity of material wasn't considered, and the centroid of cross section is a spiral line. So the slender ball screw can be regarded as the cylindrical helical spring with the great axial stiffness and tight and smooth structure on both ends, as shown in Fig. 2-2.

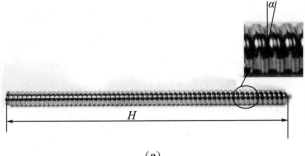

(a)

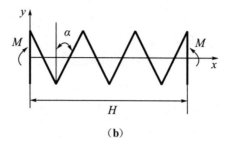

(b)

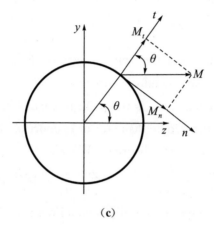

(c)

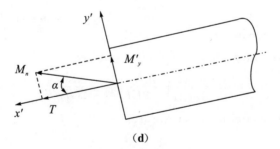

(d)

**Fig. 2-2  Equivalent model of lead screw**

The pure bending tends to take place under the action of moments. Assuming that the height of the spring is $H$, $\rho$ is the curvature radius under bending, and $\alpha$ is the helix angle of the spring, the work for the moment can be represented in the following equation:

$$U = \frac{1}{2} M \frac{H}{\rho} \qquad (2-1)$$

$M$ is the principal moment in arbitrarily section of cylindrical spring, and the radial and tangential components of $M$ are: $M_n = M\sin\theta$, $M_t = M\cos\theta$. $M_n$ can be divided into torque $T$ and bend moment $M'_y$ which is vertical or tangent to cross-section of spring wire.

$$\left.\begin{array}{l} M'_y = M\sin\theta\sin\alpha \\ T = M\sin\theta\cos\alpha \end{array}\right\} \qquad (2-2)$$

By synthesizing $M'_y$ and $M_t$, the following equation can be obtained

$$M_z = \sqrt{M^2 \sin^2\alpha \sin^2\theta + M^2 \cos^2\theta} \qquad (2-3)$$

The total deformed energy can be calculated as follows:

$$W = \int_0^{2n\pi} \left( \frac{\sin^2\alpha \sin^2\theta + \cos^2\theta}{2EI_z} + \frac{\cos^2\alpha \sin^2\theta}{2GI_P} \right) \frac{M^2 D}{2\cos\alpha} d\theta \qquad (2-4)$$

where, $n$ is the number of active coils of spring; $E$ is the elastic modulus of materials; $G$ is the shear modulus of materials; $I_z$ is the inertia moment of cross section of spring wire; $I_P$ is the polar moment of inertia of cross section of spring wire; $D$ is the middle diameter of cylindrical spring.

According to work-energy theorem, $W = U$.

$$\int_0^{2n\pi} \left( \frac{\sin^2\alpha \sin^2\theta + \cos^2\theta}{2EI_z} + \frac{\cos^2\alpha \sin^2\theta}{2GI_P} \right) \frac{MD}{\cos\alpha} d\theta = \frac{H}{\rho} \qquad (2-5)$$

The relationship between helical pitch and helix angle is taken as

$$H = n\pi D \tan\alpha \qquad (2-6)$$

$$\frac{1}{\rho} = \frac{M}{n\pi \sin\alpha} \int_0^{2n\pi} \left( \frac{\sin^2\alpha \sin^2\theta + \cos^2\theta}{2EI_z} + \frac{\cos^2\alpha \sin^2\theta}{2GI_P} \right) d\theta \qquad (2-7)$$

The equation integrating (2-7) is given by

$$\frac{1}{\rho} = \frac{M}{\sin\alpha} \left( \frac{\sin^2\alpha + 1}{2EI_z} + \frac{\cos^2\alpha}{2GI_P} \right) \qquad (2-8)$$

Based on the definition of curvature:

$$\frac{1}{\rho} = \frac{d^2 y}{dx^2} \qquad (2-9)$$

Therefore

$$\frac{d^4 y}{dx^4} = \frac{q}{\sin\alpha} \left( \frac{\sin^2\alpha + 1}{2EI_z} + \frac{\cos^2\alpha}{2GI_P} \right) \qquad (2-10)$$

where, $q$ is transverse distributed load.

When

$$q = -\frac{A\gamma l}{H} \cdot \frac{\partial^2 y}{\partial t^2} \qquad (2-11)$$

## Chapter 2  The Vibration Modal Analysis of the Ball Screw about Machine Tools

where, $\gamma$ is mass per unit volume.

$$\frac{\partial^4 y}{\partial x^4} = -\frac{A\gamma l}{H} \cdot \frac{1}{\sin\alpha}\left(\frac{\sin^2\alpha + 1}{2EI_z} + \frac{\cos^2\alpha}{2GI_P}\right)\frac{\partial^2 y}{\partial t^2} \qquad (2-12)$$

Assume that

$$a^2 = \frac{H\sin\alpha}{A\gamma l\left(\dfrac{\sin^2\alpha + 1}{2EI_z} + \dfrac{\cos^2\alpha}{2GI_P}\right)} \qquad (2-13)$$

Therefore

$$a^2 \frac{\partial^4 y}{\partial x^4} + \frac{\partial^2 y}{\partial t^2} = 0 \qquad (2-14)$$

The equation (2−14) is the partial differential equation of free vibration about equivalent beam.

The separation variable method is adopted to solve it. Because the vibration mode has nothing to do with the time, the general solution is assumed as follows:

$$y(x,t) = Y(x) \cdot Z(t) \qquad (2-15)$$

The modal can be obtained by substituting equation (2−15) into equation (2−14).

$$a^2 \frac{d^4 Y}{dx^4} \cdot \frac{1}{Y} = -\frac{d^2 Z}{dt^2} \cdot \frac{1}{Z} \qquad (2-16)$$

The left and right ends of equation (2−14) are constant which is assumed to $p^2$.

$$\frac{d^2 Z}{dt^2} + p^2 Z = 0 \qquad (2-17)$$

$$\frac{d^4 Y}{dx^4} - k^4 Y = 0 \qquad (2-18)$$

where, $k^4 = \dfrac{p^2}{a^2}$.

The solution of the vibration equation is obtained by equations (2−17) and (2−18).

$$y(x,t) = (C_1 \sin kx + D_1 \cos kx + C_2 \sinh kx + D_2 \cosh kx)[A\sin\omega t + B\cos\omega t] \qquad (2-19)$$

where, $A$, $B$, $C_1$, $C_2$, $D_1$ and $D_2$ are coefficients and determined by boundary conditions.

While

Therefore
$$x=0$$
$$y = 0, \frac{\partial y}{\partial x} = 0 \qquad (2-20)$$

$$C_1 \sin kx + D_1 \cos kx + C_2 \sinh kx + D_2 \cosh kx = 0$$

While
$$x = H$$

Therefore
$$y = 0, \frac{\partial^2 y}{\partial x^2} = 0 \qquad (2-21)$$

Therefore
$$D_1 = -D_2, \quad C_1 = -C_2$$

By equation (2−21), the equation of frequency can be derived.
$$\tan kH = \tanh kH \qquad (2-22)$$

The solution of equation can be obtained by MATLAB.
$k_1 H = 3.9266, \ k_2 H = 7.0686, \ k_3 H = 10.2102, \ k_4 H = 13.3518$

## 2.4 Analysis of speed ball screw based on FEM

The entity model of single-nut ball screw for SFU2005−800 is established and imported into ANSYS. The model material is given for GCr15. The screws are divided by adopting the structural unit with 3D 10-node tetrahedron and smart grid meshed. The meshing refinement level is 5, and 29498 units and 54672 nodes are obtained, as shown in Fig. 2−3. To simplify the analysis, the Lanczos Block method is used to solve the model under the free boundary condition. Because the external excitation (noise) is low frequency commonly, the first seven order vibration modes are analysed mainly, as shown in Fig. 2−4.

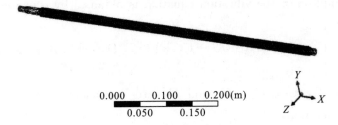

Fig. 2−3  Finite element model of a ball screw

Chapter 2　The Vibration Modal Analysis of the Ball Screw about Machine Tools

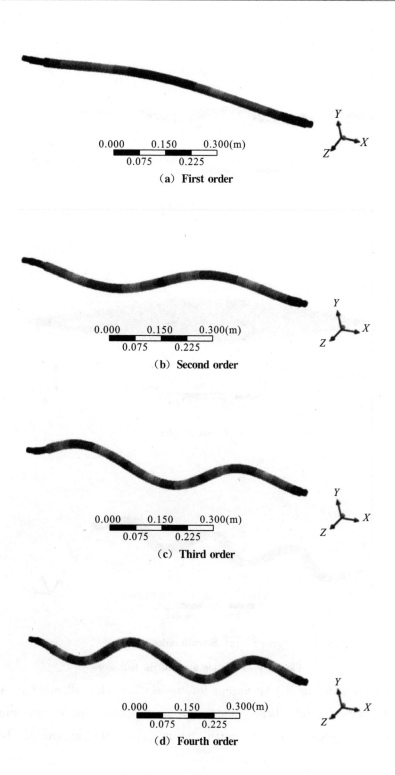

(a) First order

(b) Second order

(c) Third order

(d) Fourth order

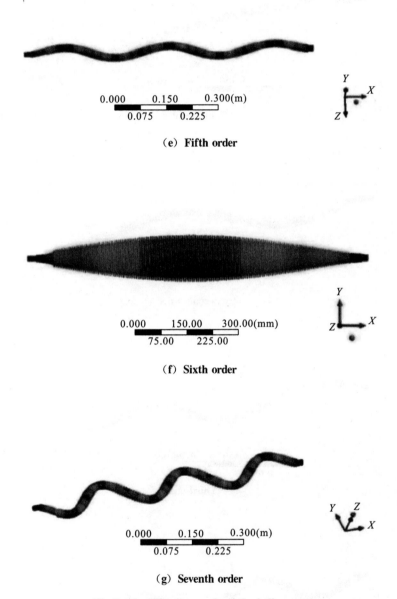

(e) Fifth order

(f) Sixth order

(g) Seventh order

Fig. 2−4  Vibration mode of the ball screw

As shown in Fig. 2−4, except for the sixth mode, all other modes are bending modes, and these modes are orthogonal modes in mutually perpendicular planes. It is worth noting that the maximum bending deformation of each mode is near the middle of the screw thread. However, the torsional expansion mode appears in the sixth vibration mode of the screw, as shown in Fig. 2-4(f), and the natural frequency value is 1757.36 Hz, which means that if there is resonance between the external excitation and the screw at the frequency, the screw will have serious torsional expansion deformation, even causing screw damage, and the maximum deformation is in the middle of the screw.

## 2.5 Modal experiment of ball screw

### 2.5.1 Modal experiment scheme

Ideally, the modal parameters of the specimen should be tested under the working condition. However, it is not easy to measure the ball screw at high speed, so the experiment is a modal parameter test under static support. The single point excitation method is used to verify the results of the finite element analysis. The DH5922N modal test system (Jiangsu Donghua Test Co., Ltd.) is used, and the composition is shown in Fig. 2-5.

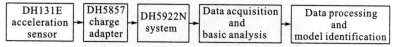

Fig. 2-5  Composition of modal analysis system

The test plan is as follows:

(1) The support mode is fixed at both ends of the screw. In order to prevent the support's inherent vibration characteristics from affecting the modal parameter test of ball screw, the support should be as thick as possible. Because the support is large relative to the screw, the support is regarded as rigid approximately, and its influence on the modal parameters of the screw is not considered[11-13].

(2) Determination of mode order. The external excitation is generally low frequency component, far away from the working frequency band of the screw, so the experiment mainly detects the natural frequency of the first four modes[14-16].

(3) Determination of measuring point position. The SFU 2005-800 ball screw is divided into 11 parts equally and 10 measuring points are arranged.

(4) Determination of incentive plan. ①The location of excitation point should avoid the expense of bearing point and structural modal node, so as to ensure the high signal-to-noise ratio of signal at measured point and avoid modal omission. ②The location of excitation point which is convenient for energy transmission should be selected. Generally, the stiffness of the location should be as large as possible. The excitation point is the ideal excitation position at 3/8 of the screw, and the frequency range is 0~2 kHz.

(5) DH131E sensor is adopted, and installed with beeswax.

On the experimental platform, the screw is hammered with pulse excitation, and the frequency response function of each measuring point on the screw is calculated through analysis software. The frequency response function of each measuring point is imported into the modal analysis system for parameter identification, and then the modal parameters such as natural frequency[17], vibration mode and damping of the screw can be estimated. The measurement experiment is shown in Fig. 2−6.

Fig. 2−6  Experimental platform for modal measurement of screw

## 2.5.2  Analysis of experimental results

The first four natural frequencies of SFU 2005−800 screw are measured in modal analysis experiment. The superposition diagram of frequency response function at each measuring point is shown in Fig. 2−7.

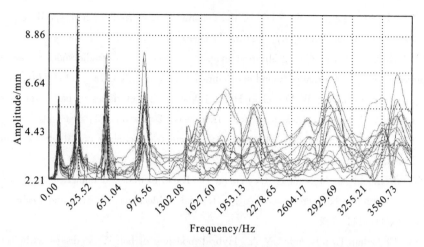

Fig. 2−7  Superposition diagram of frequency response function

The first four natural frequencies of the screw obtained through theoretical calculation, finite element simulation and actual measurement are compared, as shown in Table 2−1. It can be seen that compared with the theoretical calculation value of the natural frequency of the screw, the error of each mode is less than 5%, which shows that the mathematical model established is accurate and reliable.

Table 2−1  Comparison of natural frequencies of theoretical, simulation and experimental modes

| Name | First order | Second order | Third order | Fourth order |
| --- | --- | --- | --- | --- |
| Experimental mode/Hz | 104.98 | 306.80 | 618.49 | 1029.46 |
| Simulation mode/Hz | 108.21 | 322.46 | 638.05 | 1058.12 |
| Theoretical mode/Hz | 107.56 | 321.81 | 640.12 | 1074.93 |
| Maximum error/% | 3.08 | 4.95 | 3.49 | 4.41 |

# References

[1] Zhou Y, Wang G X, Cao X H. A measuring and analysis method to identify torsional vibration modes of ball screw feed drives [J]. China Mechanical Engineering, 2013, 24 (23): 3240-3244.

[2] Li B, Tu Z M, Mao K M. Model test and dynamic modeling of ball screw using finite element method [J]. Journal of Huazhong University of Science and Technology (Natural Science Edition), 2013, 41 (8): 74-78.

[3] Mu S G, Feng X Y. Study of the dynamic characteristic of high-speed ball screw [J]. Journal of Hunan University (Natural Science), 2011, 38 (12): 25-29.

[4] Gu U C, Cheng C C. Vibration analysis of a high-speed spindle under the action of a moving mass [J]. Journal of Sound and Vibration, 2004, 4-5 (278): 1131-1146.

[5] Yang Y, Zhang W M, Zhao H P. Dynamic characteristics of ball screw system [J]. Journal of Vibration, Measurement & Diagnosis, 2013, 33 (8): 664-669.

[6] Dadalau A, Mottahedi M, Groh K. Parametric modeling of ball screw spindles [J]. Production Engineering, 2010, 4 (6): 625-631.

[7] Romuald B, Mohammed C, Yoann B. Experimental and numerical study of the load distribution in a ball-screw system [J]. Journal of Mechanical Science and Technology, 2014, 28 (4): 1411-1420.

[8] Chinedum E, Okwudire Y A. Hybrid modeling of ball screw drives with coupled axial, torsional, and lateral dynamics [J]. Journal of Mechanical Science and Technology, 2009, 131 (7): 10021-10029.

[9] Xu H H, Li X Y. Modal analysis of a grinding wheel carriage for a vertical glass-edge grinding machine with a long ball screw [J]. Journal of Vibration and Shock, 2013, 32 (18): 189-194.

[10] Xia J Y, Hu Y M, Wu B. Research on thermal dynamics characteristics and modeling approach of ball screw [J]. The International Journal of Advanced Manufacturing Technology, 2009, 43 (5-6): 421-430.

[11] Lin J L. The modal analysis and experiment [M]. Beijing: Tsinghua University Press, 2011.

[12] Jiang B H, Qi Q, Shao R Y. Analysis of modal characteristics of test platform for ball screw pair based on ABAQUS [J]. Machine Tool & Hydraulics, 2019, 47 (17): 183-186.

[13] Fu Z B, Wang T Y, Zhang L, et al. Dynamic modeling and vibration analysis of ball screw feed driving systems [J]. Journal of Vibration and Shock, 2019, 38 (16): 56-63.

[14] Chen G Q, Li X F. Modal analysis on the linear module with ball screw [J]. Journal of Machine Design, 2019, 36 (3): 56-61.

[15] Gao D Q, Tian Z Y, Hao D J, et al. Dynamic characteristics analysis of leading screw based on ANSYS software [J]. Hoisting and Conveying Machinery, 2008 (11): 54-57.

[16] Ansoategui I, Campa F J. Mechatronics of a ball screw drive using an N degrees of freedom dynamic model [J]. International Journal of Advanced Manufacturing Technology, 2017, 93 (1-4): 1307-1318.

[17] Rinchi M, Gambini E. Theoretical and numerical experiences on a test rig for active vibration control of mechanical systems with moving constraints [J]. Shock and Vibration, 2004, 11 (3-4): 187-197.

# Chapter 3  Rigid-flexible Coupling Model and Dynamic Characteristic Analysis of Feed System

## 3.1  Introduction

In the existing research, the feed system is generally considered both as a rigid body and a flexible body[1-4]. As a flexible body, the number of degrees of freedom is large and the degree of freedom calculation is complicated[5-6]. However, as a rigid body, it conflicts with the slender lead screw[7]. In this chapter, based on the modal analysis of lead screw and the problems existing in the vibration analysis of the feed system, the rigid-flexible coupling method is adopted to analyze the relevant factors that affect the axial vibration of the feed system, and to find the most important factors. What's more, the research content of this part is the theoretical basis for the quantitative analysis of the feed system in the later stage.

## 3.2  Mathematical modeling

### 3.2.1  Dynamic model of linear guide pair

Fig. 3-1 is the vibration diagram of the linear guide rail system in the $y$-direction. The oil film between the slide and the guide is simplified into a nonlinear spring damping unit. $m$ is the total mass of the slide and the workbench, $k$ is the oil film stiffness, and $c$ is the oil film damping. The vibration equation of the system[8] in this direction is

$$m\ddot{y} + c\dot{y} + ky = 0 \tag{3-1}$$

Similarly, the vibration equation in the $x$-direction is

$$m\ddot{x} + c\dot{x} + kx = 0 \tag{3-2}$$

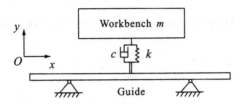

Fig. 3—1  Calculation model of linear guide pair

### 3.2.2  Dynamic model of ball screw

The vibration of the ball screw mainly includes transverse vibration, axial vibration, and torsional vibration, which has an important impact on the vibration of the workbench feeding process.

(1) The dynamic model of transverse vibration: the ball screw is equivalent to the Timoshenko beam model with elastic support at both ends, uniform material, and equal cross-section[9]. The deformation coordination condition[7] is as follows:

$$\theta = \frac{\partial v}{\partial x} + \frac{Q}{\mu AG} \tag{3-3}$$

$$M = EI \frac{\partial \theta}{\partial x} \tag{3-4}$$

where, $x$ is the axial length of the lead screw; $v$ is the transverse deformation of a lead screw; $Q$ is the shear force in cross-section of the screw; $\mu$ is the section shear coefficient; $G$ is the shear modulus; $A$ is the cross-sectional area of the screw; $E$ is the elasticity modulus; $I$ is the inertial moment; $\theta$ is the normal angle.

Then, the dynamic equilibrium equation of the screw in the vertical direction is

$$\rho A \frac{\partial^2 v}{\partial t^2} + \frac{\partial}{\partial x}\left[\mu AG\left(\theta - \frac{\partial v}{\partial x}\right)\right] = 0 \tag{3-5}$$

$$\rho I \frac{\partial^2 \theta}{\partial t^2} - \frac{\partial}{\partial x}\left(EI \frac{\partial \theta}{\partial x}\right) + \mu AG\left(\theta - \frac{\partial v}{\partial x}\right) = 0 \tag{3-6}$$

where, $\rho$ is the mass of lead screw per unit length.

By eliminating $\theta$, the horizontal free vibration equation of the ball screw can be obtained:

$$\left(EI\frac{\partial^4 v}{\partial x^4}+\rho A\frac{\partial^2 v}{\partial t^2}\right)-\frac{EI\rho}{\mu G}\cdot\frac{\partial^4 v}{\partial x^2\partial t^2}-\rho I\frac{\partial^4 v}{\partial x^2\partial t^2}+\frac{\rho^2 I}{\mu G}\cdot\frac{\partial^4 v}{\partial t^4}=0 \quad (3-7)$$

The first item in equation (3-5) is the basic item, and the second item is the effect of shear deformation. Item 3 is the effect of the moment of inertia, and item 4 is the comprehensive effect of shear and moment of inertia[10].

(2) Axial vibration equation: the axial vibration model of the ball screw can be simplified as an equal-section axial tension and compression bar with simple support at both ends[11], and its coordinate system is shown in Fig. 3-1. It is assumed that $EA$ is the tensile (compressive) stiffness of the screw, the mass per unit length in the $x$-direction is $m$, and the displacement in the $x$-direction is $u(x, t)$ as a function of the coordinate $y$ and time $t$.

According to the relationship among the Darembery's principle, axial force, and axial deformation, it can be concluded that

$$m\frac{\partial^2 u(x,t)}{\partial t^2}-EA\frac{\partial u^2(x,t)}{\partial x^2}=0 \quad (3-8)$$

(3) Torsional vibration model: The torsional vibration model of the ball screw can be simplified as a slender rod, and assuming that $\beta$ is the torsion angle, its vibration equation can be obtained from the deformation coordination condition:

$$\rho\frac{\partial^2 \beta}{\partial t^2}=G\frac{\partial^2 \beta}{\partial z^2} \quad (3-9)$$

## 3.3　Establishment of rigid-flexible coupling model

### 3.3.1　Establishment of the flexible body model

The large-scale dynamics analysis software ADAMS is adopted as a simulation tool. There are two ways to establish a flexible body in the software: ① The third-party finite element analysis software, such as ANSYS, is used to convert it into .mnf modal neutral file, and ADAMS software is used to replace the rigid body[12]. ②By adopting the flexibility function of ADAMS software, the rigid body is directly converted into the flexible body. As the ball screw is a long and thin part with large deformations, it is not consistent with the actual situation to treat it simply as a rigid body. In this book, ANSYS software is used to establish the three-dimensional model of ball screw and define its material properties and two end constraints. In addition, the beam 189 unit is used for meshing to facilitate the calculation of the Timoshenko beam model in the future[13-14].

### 3.3.2 Rigid body model establishment and simulation parameter setting

First, the rigid body in the feed system is modeled by PROE software, and imported into the ADAMS software via the .X_T file, and the component material properties and constraint relationships are set. And then, the lead screw and the nut are connected by a spiral pair with a lead of 5 mm. Next, the contact constraint is adopted between the guide rail and the workbench. The damping coefficient is initially set to 50 N · s · mm$^{-1}$. The stiffness is set to $1 \times 10^5$ N/mm. Besides, the Coulomb friction setting is adopted, and the setting, in which the static friction coefficient is 0.3 and the dynamic friction coefficient is 0.1. The established rigid-flexible coupling model is shown in Fig. 3-2. And the vibration acceleration at the center of mass of the workbench can be obtained by setting the number of simulation steps and time, as shown in Fig. 3-3.

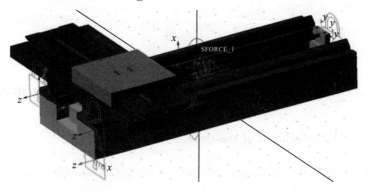

Fig. 3-2 Rigid-flexible feed system model

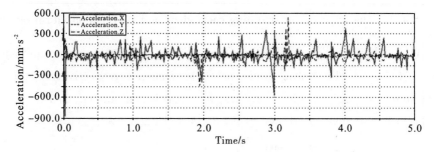

Fig. 3—3   Acceleration curve at the center of mass of the workbench

## 3.4   Analysis of simulation results

### 3.4.1   Correlation analysis

Under the initial setting conditions, the vibration acceleration curves of the workbench and the center of mass of the screw, and the $X$-axis, $Y$-axis, $Z$-axis of the linear guide are measured separately. Then output data in ADAMS/PostProcessor, and import it into EXCEL for data analysis to obtain its cross-correlation coefficient. And it can be used to analyze the relationship among the three directions vibration of the lead screw, linear guide pair, and workbench, as shown in Table 3—1.

Table 3—1   Cross-correlation coefficient of the workbench, lead screw and linear guide pair

|  | Direction | Lead screw | Linear guide pair |
|---|---|---|---|
| Workbench | $X$-axis | 0.12 | 0.79 |
|  | $Z$-axis | 0.15 | 0.82 |
|  | $Y$-axis | 0.21 | 0.69 |

As can be seen from Table 3—1, the cross-correlation coefficient between the linear guide pair and the workbench is far greater than the mutual coefficient between the lead screw and the workbench. So, it can be concluded that the vibration of the workbench is mainly caused by the vibration of the linear guide pair, and the vibration of the lead screw has little impact. Therefore, how to reduce the vibration of the linear guide pair effectively becomes the key to reduce the vibration without considering the support of the lead screw.

### 3.4.2 The influence of damping coefficient and stiffness on vibration

When the other parameters remain unchanged, the damping coefficient and stiffness of the linear guide pair are changed respectively, and the maximum vibration acceleration at the workbench under each working condition is obtained, as shown in Fig. 3−4 and Fig. 3−5.

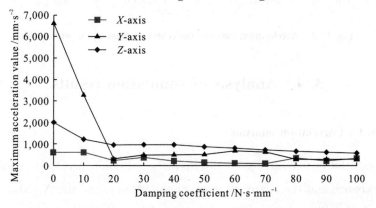

Fig. 3−4　Effect of guide rail damping on vibration acceleration

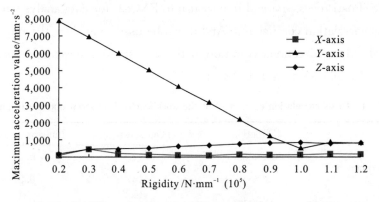

Fig. 3−5　Effect of guide rail stiffness on vibration acceleration

It can be seen from Fig. 3−4 that the influence of the guide rail damping coefficient on the maximum vibration acceleration of the workbench is the largest in the $Y$-axis and the smallest in the $X$-axis. And the maximum vibration acceleration values in all three directions decrease as the damping coefficient increases.

It can be seen from Fig. 3−5 that the stiffness of the guide rail has the greatest influence on the vibration in the $Y$-axis direction, and decreases with increasing stiffness. And the vibration in the $X$-axis and $Z$-axis directions does not change much as the stiffness increases.

Instead of the traditional method, the method of rigid-flexible coupling is adopted to analyze the dynamic characteristics of the feed system, which is closer to its real condition. In this chapter, the influence of various factors on the axial vibration is effectively analyzed, constituting the basis of the following chapters.

# References

[1] Li B, Luo B, Mao X Y. A new approach to identifying the dynamic behavior of CNC machine tools with respect to different worktable feed speeds [J]. International Journal of Machine Tools and Manufacture, 2013, 72 (9): 73-84.

[2] Edward M, Lukasz N. Vibrations in the machining system of the vertical machining center [J]. Procedia Engineering, 2012 (39): 405-413.

[3] Zhou Y, Xu X S, Wang G X, et al. Study on axial vibration of feed drives and rolling vibration of ball screw [J]. Journal of Central South University (Science and Technology), 2013, 44 (10): 4069-4073.

[4] Zhang Z Y, Song X H, Jiang H K. Experiment study on exciting factors of axial vibration of precision ball screw [J]. Journal of Vibration, Measurement & Diagnosis, 2008, 28 (1): 14-17.

[5] Zhang H D, Sun J L. The lateral vibration analysis of the drive screw under the elastic supports [J]. Journal of Changchun University, 2011, 21 (2): 16-20.

[6] Wu Q, Rui Z Y, Yang J J. Bifurcation and chaos characteristics for ball-screw system considering non-linear elastic force [J]. Journal of Xi'an Jiaotong University, 2012, 46 (1): 70-75.

[7] Hu H Y. Mechanical vibration foundation [M]. Beijing: Beihang University Press, 2005.

[8] Teng H W, Wang T, Huo D, et al. Theoretical analysis and modal test for structural damage diagnosis based on axial vibration [J]. Journal of Vibration and Shock, 2010, 29 (12): 122-125.

[9] Long W, LüB, Shi C B, et al. Torsional vibration characteristics of transmission shaft for screw drill [J]. Coal Science and Technology, 2013, 41 (8): 313-316.

[10] Xuan H, Hua Q S, Zhang H X, et al. Characteristic analysis of rigid-flexible coupling model worktable for high precision NC lathe feed system [J]. Journal of Qingdao

University (Engineering & Technology Edition), 2017, 32 (4): 68-71.

[11] Rui Z Y, Zhang G T, Feng R C, et al. Double-driven feed system model with rigid-flexible coupling and its parameter optimization [J]. Journal of Lanzhou University of Technology, 2015, 41 (4): 41-45.

[12] Wang L, Li W B. Dynamic co-simulation for servo feed system of CNC cutting machine [J]. Manufacturing Automation, 2013, 35 (8): 139-141.

[13] Luo R N, Niu W T, Wang C S. Analysis of influencing factors of dynamic error of feed system based on electromechanical-rigid-flexible coupling characteristics [J]. Chinese Journal of Engineering Design, 2019, 26 (5): 561-569.

[14] Neugebauer R, Scheffler C, Wabner M. Implementation of control elements in FEM calculations of machine tools [J]. CIRP Journal of Manufacturing Science and Technology, 2011, 4 (1): 71-79.

# Chapter 4  Static Stiffness Analysis of Feed System

## 4.1  Introduction

At present, the accuracy of ultra-precision positioning is close to the nanometer level, and the movement of nano-scale resolution has become a landmark indicator of feed system precision[1-3]. But the axial static stiffness of the feed system has a direct impact on its positioning accuracy, stability, vibration, and noise. Therefore, it is of great engineering significance to study its axial static stiffness.

The main factors affecting the axial stiffness of the ball screw feed system include axial stiffness of the screw, axial stiffness of nut, axial stiffness of bearing, axial stiffness caused by torsion of a screw, axial stiffness caused by self-weight bending of the screw, etc. Nowadays, in the traditional static stiffness model, the non-isostatic stiffness of the bearings at both ends of the screw is not considered. Generally, the series-parallel model of the feed system is established according to the different bearing types at both ends of the screw[4]. Besides, the research on the stiffness of the feed system[5-9] mainly considers the axial static stiffness caused by the deformation of the screw, nut, and bearing, and generally ignores the influence of the torsional deformation on the static stiffness. However, the dynamic performance experiment of the ball screw feed system[10-12] shows that the torsional stiffness of the ball screw also has an important influence on the dynamic characteristics of the feed system. The axial stiffness caused by the torsional deformation is neglected so that the positioning accuracy is significantly reduced. It can be seen that although the existing research has done a lot of work in the modeling of the feed system, there are still some deficiencies in the influence mechanism of the static stiffness of the system, the stiffness model under the condition of non-equal stiffness support, and most of them ignore the impact of the torsional deformation of the screw on the axial static stiffness of the system.

## 4.2　Mathematical model of axial static stiffness

### 4.2.1　A universal model of axial static stiffness of ball screw pair

In practice, the ball screw pair mainly bears the axial load, and the radial load is its weight. And in this book, the effects of radial loads are ignored. What's more, both ends of the ball screw are supported by bearings with unequal stiffness, which is a statically indeterminate problem. The mechanical model is shown in Fig. 4−1.

**Fig. 4−1　Force and support at both ends of the ball screw**

It is assumed that the axial load of the tool holder is $F$, the distance from the end face of $B$ is $x$, the total length of the lead screw is $L$, and $k_1'$ and $k_1''$ are the bearing stiffnesses at the left and right ends of the lead screw. The supporting forces $F_{Bx}$ and $F_{Cx}$ on both ends meet the static equilibrium conditions and deformation coordination conditions:

$$\begin{cases} F_{Bx} + F_{Cx} = F \\ \delta' = \delta'' \end{cases} \quad (4-1)$$

where, $\delta'$, $\delta''$ is the amount of axial deformation between the $B$ and $C$ ends to the external force acting surface.

The following formula is derived:

$$F_{Bx} = \frac{\eta}{1+\eta} F \quad (4-2)$$

$$\eta = \frac{EAk_1' + k_1'k_1''x}{EAk_1' + k_1'k_1''(L-x)} \quad (4-3)$$

by

$$\begin{cases} \delta' = \dfrac{F_{Bx}(L-x)}{EA} + \dfrac{F_{Bx}}{k_1'} \\ \delta'' = \dfrac{F_{Cx}x}{EA} + \dfrac{F_{Cx}}{k_1''} \end{cases} \quad (4-4)$$

where, $E$ is Young's modulus of elasticity of ball screw; $A$ is the cross-sectional area of the lead screw.

According to equations (4-2) and (4-3), $\delta'$ can be obtained.

From $k = dF/d\delta$, it can be obtained that the axial static stiffness of the ball screw under the support of any stiffness bearing at both ends is

$$k_1 = \dfrac{\eta(L-x)}{EA(1+\eta)} + \dfrac{\eta}{k_1'(1+\eta)} \quad (4-5)$$

It is easy to see from equation (4-5) that when $k_1' = k_1''$, calculation is consistent with the traditional static stiffness model [11].

### 4.2.2 Axial static stiffness of nut

Ignore the initial preload between the ball and the lead screw and nut. And in this book, the single nut ball screw is taken as an example to analyze its axial stiffness. The force of the ball is shown in Fig. 4-2. Furthermore, according to the Hertz contact theory, the contact deformation is as follows:

$$\delta = \delta^* \left( \dfrac{3F_n}{2E' \sum \rho} \right)^{\frac{2}{3}} \dfrac{\sum \rho}{2} \quad (4-6)$$

where, $F_n$ is the normal pressure of single ball contact point; $\sum \rho$ is the comprehensive curvature of the contact point; $E'$ is the equivalent elastic modulus of two contact objects; $\delta^*$ is the dimensionless contact deformation, which is Hertz contact coefficient.

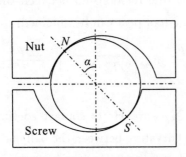
Fig. 4-2 The force of the ball

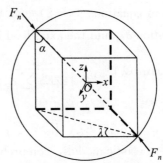
Fig. 4-3 Spatial load on ball

According to the spatial relationship in Fig. 4−3, the normal force $F_n$ that a single ball bears is

$$F_n = \frac{F}{Z\sin\alpha\cos\lambda} \qquad (4-7)$$

where, $Z$ is the number of working balls.

Under the action of normal force $F_n$, the amount of normal deformation includes two parts[13]: ①Normal deformation $\delta_{bs}^{(n)}$ between the ball and screw raceway. ②Normal deformation $\delta_{bn}^{(n)}$ between the ball and nut raceway. As shown in Fig. 4−4, normal deformation $\delta_{bs}^{(n)}$ between ball and screw raceway are

$$\delta_b^{(n)} = \delta_{bs}^{(n)} + \delta_{bn}^{(n)} \qquad (4-8)$$

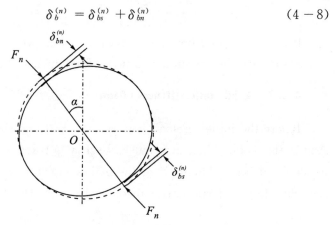

Fig. 4−4  Normal deformation of the ball

It can be seen from Fig. 4−2 that the relationship between the amount of axial deformation $\delta_{b1}$ and the amount of normal deformation $\delta_b^{(n)}$ are satisfied:

$$\delta_{b1} = \delta_b^{(n)} \sin\alpha\cos\lambda \qquad (4-9)$$

From equation (4−6), $\delta_{bs}^{(n)}$ and $\delta_{bn}^{(n)}$ can be obtained respectively, and then from equations (4−8) and (4−9), the relationship between the axial deformation $\delta_{b1}$ and the axial load $F$ of the tool post can be obtained:

$$\delta_{b1} = (\sin\alpha\cos\lambda)^{\frac{1}{3}}\left[\delta_{bs}^*\left(\frac{3}{2ZE'_{bs}\sum\rho_s}\right)^{\frac{2}{3}}\frac{\sum\rho_s}{2} + \delta_{bn}^*\left(\frac{3}{2ZE'_{bm}\sum\rho_N}\right)^{\frac{2}{3}}\frac{\sum\rho_N}{2}\right]F^{\frac{2}{3}}$$

$$= k_{n1}F^{\frac{2}{3}} \qquad (4-10)$$

where, $\delta_{bs}^*$, $\delta_{bn}^*$, $E'_{bs}$, $E'_{bm}$, $\sum\rho_s$ and $\sum\rho_N$ are Hertz contact coefficients, equivalent elastic moduli, comprehensive curvature between the ball and screw raceways, and comprehensive curvature between the ball and nut raceways, respectively[14].

Then the axial static stiffness of the nut is

$$k_2 = \frac{dF}{d\delta} = \frac{\frac{3}{2}\delta_{b1}^{\frac{1}{2}}}{k_{n1}} \quad (4-11)$$

It can be seen from equation (4-11) that the axial contact stiffness of the nut is closely related to the normal contact deformation of the ball. And in the process of feed system movement, its load is generally changing. Then the contact deformation changes accordingly, so the ball screw feed system is a typical non-linear system with variable stiffness.

### 4.2.3 Axial static stiffness of angular contact ball bearings

Taking a single-row angular contact ball bearing as an example, the axial stiffness analysis method is similar to that of a nut. The structural parameters of the screw and nut are replaced by the structural parameters of the bearing's inner and outer rings, respectively. There is no helix angle in the bearing. The helix angle $\lambda=0$ can be set. And then the specific analysis process is not repeated here. Its axial stiffness is

$$k_3 = \frac{dF}{d\delta} = \frac{\frac{3}{2}\delta_{b2}^{\frac{1}{2}}}{k_{n2}} \quad (4-12)$$

where, $\delta_{b2}$ is the axial deformation of bearing; $k_{n2}$ is the correlation coefficient of bearing axial stiffness.

### 4.2.4 Axial static stiffness caused by torsional deformation of the screw

The screw is regarded as an equal-length straight bar with torsional deformation. When the nut is at the $x$ position, the axial deformation $\delta_3$ is

$$\delta_3 = \frac{TL_0 x}{2\pi GI_P} \quad (4-13)$$

where, $I_P$ is the polar moment of inertia of lead screw; $G$ is the shear modulus of lead screw material; $L_0$ is the lead screw lead.

Ignoring the influence of friction factors, the axial stiffness caused by torsional deformation is

$$k_4 = \frac{4\pi^2 GI_P}{L_0^2 x} \quad (4-14)$$

It can be seen that, with the movement of the table, the torsional stiffness of the lead screw changes in a non-linear pattern of "large to small", which is a characteristic of soft spring.

The equivalent axial stiffness $k$ of the feed system can be regarded as the series sum of the stiffness of the parts related to the ball screw pair, which can be obtained by equation (4−15).

$$\frac{1}{k} = \frac{1}{k_1} + \frac{1}{k_2} + \frac{1}{k_4} \qquad (4-15)$$

## 4.3 Validation of axial static stiffness model

Take the feed system test-bench shown in Fig. 4−5 as the research object, and use Matlab to solve the axial static stiffness model of the feed system. Then according to the actual measurement and calculation, the main simulation parameters of the moving parts of the feed system of the machine tool are shown in Table 4−1.

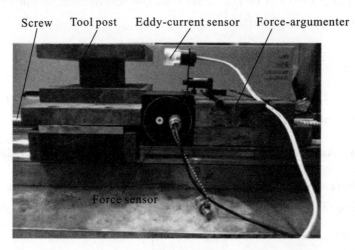

Fig. 4−5  Axial static stiffness test

The test bench uses TBI's SFU 2005 ball screw with an effective stroke of 800 mm. One end is a 7002C/P5 angular contact ball bearing, and the other end is a 6202 deep groove ball bearing. Also, a screw jack is used to apply the axial load to simulate the force of the tool post. And two force sensors measure the loading force. The eddy current sensor measures the axial displacement of the feed system. Considering that the feed process of CNC machine tools is mostly unidirectional force, the static stiffness experiment of one-way loading is carried out in the experiment, that is, the leftmost end (coordinate origin) is taken as the starting point, and the 10 kN load is applied to the feed system at the interval of 100 mm. When measuring the amount of axial deformation, to eliminate the influence of assembly and other factors on the clearance, the 30 N load is taken as the starting point of deformation measurement. To ensure the accuracy of the test results, the measurement is repeated 5 times under the same displacement. What's more, the experimental data were calculated from the average of 5 measurements.

When the feed rate is different, the measured value of the axial static stiffness of the ball screw feed system is compared with the theoretical value, as shown in Fig. 4-6. It can be seen that the theoretically calculated value of the axial static stiffness of the ball screw feed system is in good agreement with the experimentally measured value, and the maximum error is less than 3%. Compared with the reference, the error between them is obviously reduced. It can be seen that the model can accurately reflect the static stiffness of the feed system.

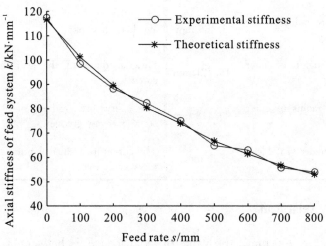

Fig. 4-6 **Comparison of measured stiffness and theoretical stiffness**

## 4.4 Influence of various factors on axial static stiffness

The Poisson's ratio of the ball screw pair and bearing material is 0.3, the elastic modulus is 206 GPa, and $F = 10$ kN. Other structural parameters are shown in Table 4−1.

Table 4−1  Structural parameters of the feed system

| Parameter | Value | Parameter | Value |
| --- | --- | --- | --- |
| Bearing contact angle | 15° | The contact angle of the lead screw | 40° |
| Number of bearing balls | 11 | Ball diameter | 3.175 mm |
| Bearing inner diameter | 15 mm | Lead screw lead | 5 mm |
| Bearing outer diameter | 32 mm | The helix angle of the lead screw | 9.087° |
| Bearing ball diameter | 4.698 mm | Number of working balls | 70 |
| The curvature radius of the inner raceway | 2.419 mm | Curvature ratio of the raceway | 1.070 |
| The curvature radius of the outer raceway | 2.466 mm | Workbench stroke | 800 mm |
| Great circle radius of inner ring raceway | 15.721 mm | Shear modulus of the lead screw | 80.77 GPa |
| Great circle radius of outer ring raceway | 20.418 mm | The curvature radius of screw raceway groove | 1.699 mm |
| Nominal diameter of the lead screw | 20 mm | The curvature radius of nut raceway groove | 1.699 mm |

### 4.4.1 Axial load

Taking $x = 400$ mm, the value of axial load $F$ is between 1 kN and 100 kN, and the relationship between axial stiffness and axial load of the feed system can be obtained from equation (4−1), as shown in Fig. 4−7.

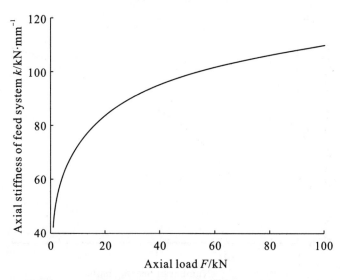

Fig. 4—7  Relationship between axial stiffness and axial load

It can be seen in Fig. 4−7 that the axial stiffness of the feed system increases with the increase of the axial load and the relationship between them is non-linear. When the axial load $F<20$ kN, the change rate of axial stiffness $k$ is high; when $F>20$ kN, the change of axial stiffness $k$ tends to be gentle.

### 4.4.2 Number of balls

It can be seen from Fig. 4−8 that the number of working balls in the bearing or lead screw will affect the axial stiffness of the feed system. The axial stiffness $k$ of the system increases significantly non-linearly with the increase of the number of working balls of the bearing. And the increase of the number of working balls of the lead screw causes the increase of the system's axial stiffness $k$ to change in an almost linear way. Therefore, for the latter, when the number of working balls increases within a certain range, the system stiffness can be regarded as a linear change.

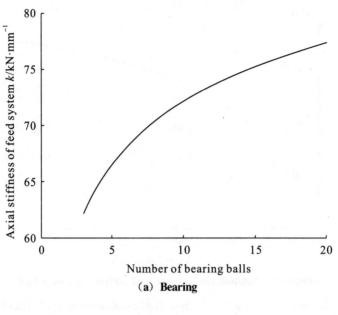

(a) Bearing

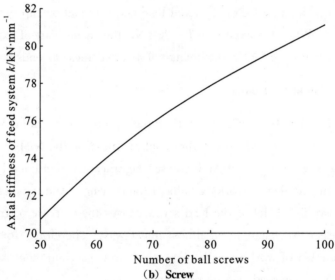

(b) Screw

Fig. 4—8 Relationship between stiffness and the number of balls

### 4.4.3 Contact angle

With the axial load of 10 kN, the bearing contact angle $\alpha'$ is selected to be $0°\sim20°$ and the screw contact angle $\alpha$ is selected to be $0°\sim50°$ for analysis.

As shown in Fig. 4 − 9, the axial stiffness $k$ of the system increases non-linearly with the increase of the screw contact angle $\alpha$. Furthermore, in the range of $\alpha > 8°$, the axial stiffness $k$ gradually changes. But with the increase of bearing contact angle $\alpha'$, the axial stiffness $k$ of the system decreases non-linearly. It is obvious that increasing the contact angle of the ball screw and bearing in the feed system can increase the axial bearing capacity of the system. However, its effect on the axial stiffness is quite the opposite.

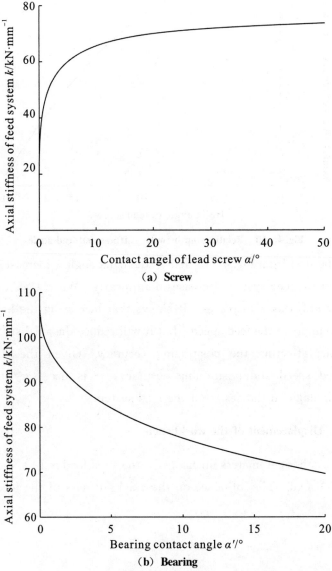

Fig. 4-9 Relationship between stiffness and contact angle

### 4.4.4 Helix angle of the lead screw

The axial load is taken as 10 kN, and the value range of helix angle $\lambda$ is $0° \sim 20°$. The analysis results are shown in Fig. 4—10.

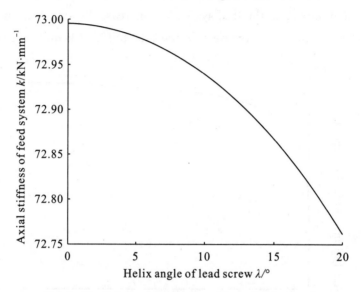

Fig. 4—10 Relationship between stiffness and lead-angle

It can be seen from Fig. 4—10 that as the helix angle $\lambda$ increases, the axial stiffness $k$ of the feed system decreases non-linearly. When $\lambda > 5°$, the change rate of axial stiffness $k$ increases. It shows that increasing the helix angle is beneficial to improve the feed speed, but it will reduce the axial stiffness of the system, thus affecting the positioning accuracy. Given the contradiction between feed speed and positioning accuracy, it is necessary to find the optimal helix angle in the design of the feed system.

### 4.4.5 Displacement of the workbench

Keeping other parameters unchanged, the axial load is 10 kN and the feed rate is $0 \sim 800$ mm. The influence on the axial stiffness of the feed system is shown in Fig. 4—11.

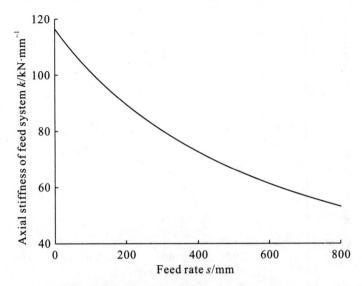

**Fig. 4—11  Relationship between stiffness and displacement of the table**

As is shown in Fig. 4 − 11, as the displacement increases, the axial stiffness of the system decreases in a non-linear manner. When the displacement increases from 0 mm to 800 mm, the axial stiffness $k$ decreases from 116.360 kN/mm to 53.146 kN/mm, which decreases by 54.33%. Therefore, reducing the axial displacement of the workbench can effectively improve the axial stiffness of the system.

### 4.4.6  Rigidity of each component

With an axial load of 5 kN, make

$$k' = \frac{k_1}{k_1 + k_2 + k_4} \qquad (4-16)$$

Set the table displacement $s$ range to 0~800 mm, and the relationship between $k'$ and $s$ is shown in Fig. 4—12 (a).

It can be seen from Fig. 4—12 (a) that during the feeding process, the value of $k'$ becomes approximately parabolic. When $s=0$ mm, $k'=99.92\%$; when $s=800$ mm, $k'=100\%$; when $s=329$ mm, $k'_{min}=99.79\%$. It can be seen that the axial rigidity $k_1$ of the screw is far greater than $k_2$, $k_3$, and $k_4$. And from equation (4−1), it can be seen that axial rigidity of the serew has little effect on the total stiffness $k$ of the system and can be ignored.

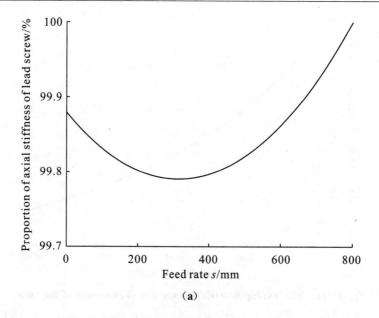

(a)

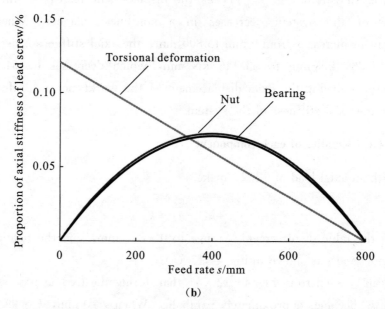

(b)

**Fig. 4−12　Relationship for the stiffness between the component and the feed system**

Similarly, the relationship between the ratio of bearing stiffness, nut stiffness and axial stiffness caused by screw torsion and the displacement of the workbench is shown in Fig. 4−12(b). It is shown that in the feed range of 0~800 mm, the axial stiffness caused by the torsion of the lead screw decreases linearly, and the stiffness of the nut and the bearing presents an approximately parabolic law. It is known from equation (4-1) that in the range of 0～332.2 mm, the effect of the stiffness of nut and the bearing on the axial stiffness of the system is the greatest; in the range of 332.2～800 mm, the axial stiffness caused by torsional deformation has the greatest impact on the axial stiffness of the system.

# References

[1] Han L, Zhang D W, Tian Y L, et al. Static stiffness modeling and sensitivity analysis for geared system used for rotary feeding [J]. Proceedings of the Institution of Mechanical Engineers Part C—Journal of Mechanical Engineering Science, 2014, 228 (8): 1431-1443.

[2] Verma K, Belokar R M. Inclusive estimations of ball screw-based CNC feed drive system over positioning and pre-loading factor [J]. Assembly Automation, 2018, 38 (3): 303-313.

[3] Ohashi T, Shibata H, Futami S, et al. Influence of linear ball guide preloads and retainers on the microscopic motions of a feed-drive system [J]. Journal of Advanced Mechanical Design Systems and Manufacturing, 2018, 12 (5): 1-10.

[4] Liu Y, Wang J S, Zhao T, et al. Stiffness model for a ball screw drive system [J]. Journal of Tsinghua University (Science and Technology), 2011, 51 (5): 601-606.

[5] Liu J, Ou Y. Dynamic axial contact stiffness analysis of position preloaded ball screw mechanism [J]. Advances in Mechanical Engineering, 2019, 11 (1): 1-15.

[6] Wang W, Zhang X Y, Mei X. Research on the mechanism of free surface contour error caused by the stiffness of feed system of five-axis machine tools [J]. Journal of Mechanical Engineering, 2016, 52 (21): 146-154.

[7] Chen Y J, Tang W C, Wang J L. Influence factors on stiffness of a ball screw [J]. Journal of Vibration and Shock, 2013, 32 (11): 70-74.

[8] Luo H T, Fu J, Jiao L C, et al. Theoretical calculation and simulation analysis of axial static stiffness of double-nut ball screw with heavy load and high precision [J]. Mathematical Problems in Engineering, 2019: 1-11.

[9] Wang Y L, Li Z K, Qiu J S, et al. Device and method for simultaneously measuring comprehensive static stiffness of linear guideway [J]. Journal of Huazhong University of Science and Technology (Natural Science Edition), 2017, 45 (9): 58-63.

[10] Hyuk L, Jin W S, Chong H C. Torsional displacement compensation in position

control for machining centers [J]. Control Engineering Practice, 2001, 9 (1): 79-87.

[11] Luo L, Zhang W M, Jürgen F. Excitation response characteristics of ball screw feed drive system and cosine-like jerk motion profiles [J]. Journal of Vibration, Measurement & Diagnosis, 2019, 39 (1): 160-167.

[12] Zou C F, Zhang H J, Lu D, et al. Effect of the screw - nut joint stiffness on the position-dependent dynamics of a vertical ball screw feed system without counterweight [J]. Proceedings of the Institution of Mechanical Engineers Part C — Journal of Mechanical Engineering Science, 2018, 232 (15): 2599-2609.

[13] Feng H T. Dynamics anddesign basis of ball screw pair [M]. Beijing: Mechanical Industry Press, 2015.

[14] Luo J W, Luo T Y. Analysis, calculation and application of rolling bearing [M]. Beijing: Mechanical Industry Press, 2009.

# Chapter 5  Influence of Dynamic Stiffness on Vibration of Feed System

## 5.1  Introduction

The dynamic stiffness of the feed system refers to the stiffness of feed system under the action of dynamic force[1-2]. It is also the key to the vibration analysis and parameter identification of the feed system. In the traditional dynamic stiffness model, the rigid body lumped parameter method was used instead of considering the dynamic elastic characteristics of the screw, which limited its description of the dynamic performance of the high-speed screw. In elastic dynamics modeling, finite element method was widely used[3], but its application in motion control and model identification was limited by huge matrix quantity and calculation error[4-6]. Ansoategui and Campa[7] adopted Newton's formulation to establish the lumped-parameter dynamic structural model of the feed system in modal coordinates, and compared it with the model in natural coordinates. Although it proved the validity of the model in modal coordinate system, the study only analyzed the related vibration characteristics under torsional modes. Zou et al.[8] studied the influence of screw-nut stiffness on the vibration characteristics of the spindle system in the transmission direction. Based on the equivalent dynamic equation and the D'Alembert's principle, the varying-coefficients model of feed system was established, but the influence of coupling characteristics was neglected. Vicente et al.[9] established a mathematical model, which is coupled with torsional and axial system by using Ritz series method, to analyze the influence of lead and workbench position. The coupled and non-coupled model frequencies are analyzed and compared, but the influence is neglected. Meanwhile, the torsion of the workbench during modeling did not reach a certain effect trend. Weng et al.[10] used ANSYS to establish a finite element model considering joints, and analyzed the effects of various joints stiffness on longitudinal vibration dynamic performance through modal analysis and harmonic response analysis. However, they did not establish an effective mathematical model that was only analyzed by software, which in turn comprehensively and quantitatively analyzed various factors. To study the dynamic characteristics of one-axle stage under heavy load, an equivalent finite element model of one-axle stage under heavy load was proposed, but the boundary conditions of bolted connections were ignored in the modeling. Therefore, to reduce the vibration in the feed, it is important to perform a full study taking into account all the possible factors affecting the vibration of the system[11].

## 5.2 Dynamic model considering coupling stiffness

The structure of the feed system is simplified as follows in order to establish a better mechanical model:

(1) The motor shaft is short and thick, and its rigidity is large, so it is considered as a rigid body.

(2) The coupling is simplified into an equivalent torsional spring consisting of torsional stiffness and torsional damping.

(3) Each joint is simplified into a damper consisting of the axial stiffness and axial damping along the $x$ direction and torsional stiffness and torsional damping around the $x$ direction.

(4) The workbench-guideway joint is simplified into a damper composed of damping and stiffness along the $z$ direction.

(5) The screw is considered as an elastic body.

(6) The workbench is regarded as a centralized mass.

The feed system model can be simplified as is shown in Fig. 5−1.

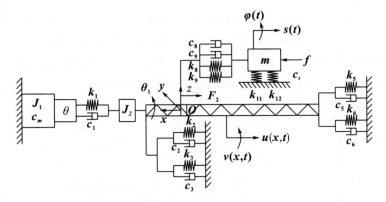

Fig. 5-1  The simplified model of the feed system

As shown in Fig. 5-1, $\theta$ and $\theta_1$ represent the input angular displacement of the motor shaft rotating around the $x$ direction and the input angular displacement of the screw beside the motor, respectively; $J_2$, $J_1$, and $J_3$ stand for the moment of inertia of coupling, the moment of inertia of motor and the moment of inertia of workbench around the $x$ direction, respectively; $c_m$ is the viscous damping of the motor around the $x$ direction; $k_1$ and $c_1$ represent torsional stiffness and the damping of coupling; $k_2$, $k_5$ and $c_2$, $c_5$ stand for the axial stiffness and the axial damping of two screw-bearing joints along the $x$ direction; $k_3$, $k_6$ and $c_3$, $c_6$ represent the torsional stiffness and the torsional damping of two screw-bearing joints around the $x$ direction; $k_8$, $k_9$ and $c_8$, $c_9$ stand for the axial stiffness along the $x$ direction and the torsional stiffness around the $x$ direction of the screw-nut joint, the axial damping along the $x$ direction and the torsional damping around the $x$ direction; $k_{11}$, $k_{12}$ and $c_{11}$, $c_{12}$ represent the radial stiffness and the radial damping of two workbench-guideway joints along the $z$ direction; $c_t$ is the axial viscous damping between workbench and guideway along the $x$ direction; $m$ is the mass of workbench; $u(x, t)$ and $v(x, t)$ represent the axial vibration displacement along the $x$ direction and the torsional vibration displacement around the $x$ direction of the screw, respectively; $s(t)$ and $\varphi(t)$ are the axial displacement and the torsional displacement of the workbench.

The second kind of Lagrange equation is expressed as

$$\frac{\mathrm{d}}{\mathrm{d}t}\left(\frac{\delta L}{\delta \dot{q}_i}\right) - \frac{\delta L}{\delta q_i} + \frac{\delta D}{\delta \dot{q}_i} = Q \qquad (5-1)$$

with $L = T - V$, where $T$ and $V$ are kinetic energy and potential energy. $L$, $D$, $Q$

and $q_i$ represent difference between kinetic energy and potential energy, damping energy, generalized force, and generalized coordinate of the system.

Kinetic energy of the system includes the kinetic energy of the workbench, the axial and the torsional vibration kinetic energy of the screw, etc., as shown in equation (5−2).

$$\begin{cases} T = \frac{1}{2}J_1\dot{\theta}^2 + \frac{1}{2}J_2\left[\dfrac{\dot{\theta}+\dot{\theta}_1+\dfrac{\partial v(0,t)}{\partial t}}{2}\right]^2 + \frac{1}{2}m\dot{s}^2 + \frac{1}{2}J_3\dot{\varphi}^2 + \\ \quad \frac{1}{2}\int_0^L \rho J\left[\dot{\theta}_1+\dfrac{\partial v(x,t)}{\partial t}\right]^2 dx + \frac{1}{2}\int_0^L \rho A\left[\dfrac{\partial u(x,t)}{\partial t}\right]^2 dx \\ J_3 = \dfrac{m}{12}\left(\dfrac{b}{2}\right)^2 \end{cases} \quad (5-2)$$

where, $J$, $A$ and $\rho$ are the polar moment of inertia, the cross-sectional area, the density and the diameter of screw, respectively; $b$ is the length of workbench along the $y$ direction.

The potential energy of the system includes the deformation energy of the screw and the elastic potential energy of each joint, as shown in equation (5−3).

$$V = \frac{1}{2}k_1[\theta_1 - \theta + v(0,t)]^2 + \frac{1}{2}k_2 u^2(0,t) + \frac{1}{2}k_6 v^2(L,t) + \frac{1}{2}k_5 u^2(L,t) +$$

$$\frac{1}{2}k_3 v^2(0,t) + \frac{1}{2}k_9[\varphi(t) - v(x,t)]^2 + \frac{1}{2}(k_{11}+k_{12})\left[\frac{b}{2}\varphi(t)\right]^2 +$$

$$\frac{1}{2}JG\int_0^L\left[\frac{\partial}{\partial x}v(x,t)\right]^2 dx + \frac{1}{2}k_8[s(t) - l\varphi(t) - u(x,t) - l\theta_1(t)]^2 +$$

$$\frac{1}{2}EA\int_0^L\left[\frac{\partial}{\partial x}u(x,t)\right]^2 dx \quad (5-3)$$

where, $G$, $E$, $l$ and $L$ are the modulus of Poisson, Young's modulus, lead and length of the screw, respectively.

The damping energy of the system includes energy consumption of the structural damping of the screw and damping at joints, as shown in equation (5−4).

$$D = \frac{1}{2}c_8\left[\dot{s} - l\dot{\varphi} - \frac{\partial u(x,t)}{\partial t} - l\dot{\theta}_1\right]^2 + \frac{1}{2}c_2\left[\frac{\partial u(0,t)}{\partial t}\right]^2 + \frac{1}{2}c_5\left[\frac{\partial u(L,t)}{\partial t}\right]^2 +$$

$$\frac{1}{2}c_3\left[\frac{\partial v(0,t)}{\partial t}\right]^2 + \frac{1}{2}c_6\left[\frac{\partial v(L,t)}{\partial t}\right]^2 + \frac{1}{2}(c_{11}+c_{12})\left(\frac{b}{2}\dot{\varphi}\right)^2 + \frac{1}{2}c_t\dot{s}^2 +$$

$$\frac{1}{2}c_{t2}\dot{\varphi}^2 + \frac{1}{2}c_1\left[\dot{\theta}_1 + \frac{\partial v(0,t)}{\partial t} - \dot{\theta}\right]^2 + \frac{1}{2}c_m\dot{\theta}^2 \quad (5-4)$$

where, $c_{t2}$ is the torsional stiffness of workbench.

The virtual work done by the external force of the system is described as

$$\begin{cases} \partial W = (F+f)\partial x \\ f = f_1 \text{sign}(\dot{s}) + f_2 \dot{s} \end{cases} \quad (5-5)$$

where, $F$ is the nut preload, which is calculated at 10% of the maximum dynamic load; $f$ is the resistance of the workbench; $f_1$ and $f_2$ are Coulomb friction coefficient and static friction coefficient of the workbench respectively.

Based on the hypothetical mode method, the torsional vibration displacement and the axial vibration displacement of the screw can be divided into two functions expressed by displacement and time[12-13].

$$\begin{cases} u(x,t) = \sum_{i=1}^{N} \cos\left[\dfrac{(i-1)\pi x}{L}\right] q_u(t) \\ v(x,t) = \sum_{i=1}^{N} \cos\left[\dfrac{(i-1)\pi x}{L}\right] q_v(t) \end{cases} \quad (5-6)$$

where, $q_u(t)$ and $q_v(t)$ are the axial independent coordinate and the torsional independent coordinate of the screw respectively; $N=3$.

Solving equations (5-1), (5-2), and (5-6), the dynamic equation of the feed system can be obtained as

$$M[\ddot{q}_i] + C[\dot{q}_i] + K[q_i] = F(t) \quad (5-7)$$

where, $M$, $C$, and $K$ are the inertia, damping, and the stiffness matrices, respectively; $q_i$ is the generalized coordinates; $F(t)$ is the generalized forces.

$$M = \begin{bmatrix} J_1 & 0 & 0 & 0 & 0 & 0 & 0 & 0 & 0 \\ 0 & \rho l J_2 + \dfrac{1}{4}J_2 & \dfrac{1}{4}J_2 & \dfrac{1}{4}J_2 & 0 & 0 & 0 & 0 & \dfrac{1}{4}J_2 \\ 0 & \dfrac{1}{4}J_2 & \dfrac{1}{2}\rho l J_2 + \dfrac{1}{4}J_2 & \dfrac{1}{4}J_2 & 0 & 0 & 0 & 0 & \dfrac{1}{4}J_2 \\ 0 & \dfrac{1}{4}J_2 & \dfrac{1}{4}J_2 & \rho A l & 0 & 0 & 0 & 0 & \dfrac{1}{4}J_2 \\ 0 & 0 & 0 & 0 & \dfrac{1}{2}\rho A l & 0 & 0 & 0 & 0 \\ 0 & 0 & 0 & 0 & 0 & \dfrac{1}{2}\rho A l & 0 & 0 & 0 \\ 0 & 0 & 0 & 0 & 0 & 0 & \dfrac{1}{2}\rho A l & 0 & 0 \\ 0 & 0 & 0 & 0 & 0 & 0 & 0 & m & 0 \\ 0 & \dfrac{1}{4}J_2 & \dfrac{1}{4}J_2 & \dfrac{1}{4}J_2 & 0 & 0 & 0 & 0 & J+J_2 \end{bmatrix}$$

$$\mathbf{K} = \begin{bmatrix}
k_9 - l_1^2 k_8 + \dfrac{(k_{11}+k_{12})b^2}{4} & l_1^2 k_8 - k_9 & B(-k_9 - l_1^2 k_8) & Z(-k_9 - l_1^2 k_8) \\
l_1^2 k_8 - k_9 & l_1^2 k_8 + 2k_3 + k_9 + k_1 & B(k_9 + l_1^2 k_8) + k_1 & Z(k_9 + l_1^2 k_8) + 2k_6 + k_1 \\
B(l_1^2 k_8 - k_9) & B(k_9 + l_1^2 k_8) + k_1 & B^2(l_1^2 k_8 + k_9) + k_1 & BZ(k_9 + l_1^2 k_8) + k_1 \\
Z(l_1^2 k_8 - k_9) & Z(l_1^2 k_8 + k_9) + k_1 + 2k_3 & BZ(l_1^2 k_8 + k_9) + k_1 + 2k_3 & Z^2(k_9 + l_1^2 k_8) + k_1 + 2k_3 \\
0 & l_1 k_8 & B l_1 k_8 & Z l_1 k_8 \\
l_1 k_8 & B l_1 k_8 & B^2 l_1 k_8 & B Z l_1 k_8 \\
l_1 k_8 & Z l_1 k_8 & B Z l_1 k_8 & Z^2 l_1 k_8 \\
B l_1 k_8 & Z B k_8 & \dfrac{JG\pi^2}{2L} & -B l_1 k_8 \\
Z l_1 k_8 & B k_8 & B l_1 k_8 & -Z l_1 k_8 \\
2k_2 + k_8 & 2k_5 + \dfrac{EA\pi^2}{2l} + B^2 k_8 & -l_1 k_8 & -k_1 \\
B k_8 & -B Z k_8 & Z^2 k_8 + 2k_5 + \dfrac{2EA\pi^2}{l} & 0 \\
Z k_8 + 2k_5 & -B k_8 & -Z k_8 & -k_1 \\
-k_8 & 0 & k_8 & -k_1 \\
0 & & 0 & k_1
\end{bmatrix}$$

## Chapter 5  Influence of Dynamic Stiffness on Vibration of Feed System

$$C = \begin{bmatrix}
c_9 - l_1^2 c_8 + \dfrac{(c_{11}+c_{12})b^2}{4} + c_{t2} & l_1^2 c_8 - c_9 & B(-l_1^2 c_8 - c_9) & Z(-l_1^2 c_8 - c_9) & l_1 c_8 & Bl_1 c_8 & Zl_1 c_8 & -l_1 c_9 \\
-l_1^2 c_8 - c_9 & -l_1^2 c_8 + c_9 + 2c_6 + c_1 & B(l_1^2 c_8 + c_9) + c_1 + 2c_3 & Z(l_1^2 c_8 + c_9) + 2c_6 + c_1 & l_1 c_8 & Bl_1 c_8 & Zl_1 c_8 & -l_1 c_8 \\
B(l_1^2 c_8 - c_9) & B(l_1^2 c_8 + c_9) + c_1 + c_6 & B^2(l_1^2 c_8 + c_9) + c_1 + 2c_6 & BZ(l_1^2 c_8 + c_9) + c_1 + c_6 & Bl_1 c_8 & B^2 l_1 c_8 & BZl_1 c_8 & -Bl_1 c_8 \\
Z(l_1^2 c_8 - c_9) & Z(l_1^2 c_8 + c_9) + c_1 + 2c_6 & ZB(l_1^2 c_8 + c_9) + c_1 + 2c_6 & Z^2(l_1^2 c_8 + c_6) + c_1 + 2c_3 & Zl_1 c_8 & BZl_1 c_8 & Z^2 l_1 c_8 & -Zl_1 c \\
l_1 c_8 & l_1 c_8 & Bl_1 c_8 & Zl_1 c_8 & c_5 + c_8 & c_5 + Bc_8 & c_5 + Zc_8 & -c_8 \\
Bl_1 c_8 & Bl_1 c_8 & B^2 l_1 c_8 & BZl_1 c_8 & c_5 + Bc_8 & c_2 + B^2 c_8 & c_5 + BZc_8 & -Bc_8 \\
Zl_1 c_8 & Zl_1 c_8 & BZl_1 c_8 & Z^2 l_1 c_8 & Zc_8 + c_2 & c_5 + BZc_8 & c_5 + Z^2 c_8 & -Zc_8 \\
-l_1 c_8 & -l_1 c_8 & -Bl_1 c_8 & -Zl_1 c_8 & -c_8 & -Bc_8 & -Zc_8 & c_5 + c_8 \\
0 & -c_1 & -c_1 & -c_1 & 0 & 0 & 0 & 0
\end{bmatrix}$$

In matrix $C$ and $K$, $B$ and $Z$ are shown in equations (5−8) and (5−9).

$$B = \cos \frac{\pi d}{2L_1} \qquad (5-8)$$

$$Z = \cos \frac{\pi d}{L_1} \qquad (5-9)$$

where, $L_1$ and $d$ are the length and the diameter of the screw at workbench.

From the matrix determinant, it can be concluded that the main factors affecting the dynamic performance of the system are the mass of the workbench, the moment of inertia of the rotating parts, the stiffness and damping of the components, the position of the workbench, and so on.

## 5.3 Experimental verification of dynamic model

The ball screw feed system of high-speed precision CNC lathe (Model CK6136S, Zhejiang Kaida Machine Tool Group Co., Ltd., China) is selected to verify the reliability of the dynamic model. The axial dynamic response characteristics of the feed system are tested in experiments. The dynamic equation of the axial vibration is decoupled and solved by Matlab. The main simulation parameters of the feed system obtained through measurement and calculation are shown in Table 5−1[14].

Table 5−1 Main calculating parameters of feed system

| Parameter | Value | Parameter | Value |
| --- | --- | --- | --- |
| $\rho / \text{kg} \cdot \text{m}^{-3}$ | $7.9 \times 10^3$ | $k_1 / \text{N} \cdot \text{m}^{-1}$ | $4.2 \times 10^5$ |
| $A / \text{m}^2$ | $8.1 \times 10^{-4}$ | $k_2, k_5 / \text{N} \cdot \text{m}^{-1}$ | $5.6 \times 10^5$ |
| $L / \text{m}$ | 1.5 | $k_3, k_6 / \text{N} \cdot \text{m} \cdot \text{rad}^{-1}$ | $6.9 \times 10^4$ |
| $L_1 / \text{m}$ | 0.45 | $k_8 / \text{N} \cdot \text{m}^{-1}$ | $1.0 \times 10^7$ |
| $l / \text{m}$ | $6 \times 10^{-3}$ | $k_9 / \text{N} \cdot \text{m} \cdot \text{rad}^{-1}$ | $5.9 \times 10^4$ |
| $r / \text{m}$ | $3.2 \times 10^{-2}$ | $E / \text{Gpa}$ | 206 |
| $m / \text{kg}$ | 136 | $G / \text{Gpa}$ | 78.6 |
| $J / \text{kg} \cdot \text{m}^2$ | $1.2 \times 10^{-3}$ | $b / \text{m}$ | 0.83 |
| $J_1 / \text{kg} \cdot \text{m}^2$ | $2.1 \times 10^{-3}$ | $k_{11}, k_{12} / \text{N} \cdot \text{m}^{-1}$ | $5.5 \times 10^9$ |
| $J_2 / \text{kg} \cdot \text{m}^2$ | $1.3 \times 10^{-4}$ | | |

In order to simplify the analysis, the damping is 0. In the simulation,

## Chapter 5  Influence of Dynamic Stiffness on Vibration of Feed System

the motor torque is used as input and the axial acceleration of workbench is used as output. The spectrum analysis results are shown in Fig. 5−2.

The test principle of the experiment is shown in Fig. 5−3. The vibration test device mainly contains DH5922N dynamic signal test and analysis system, DH311 piezoelectric acceleration sensor and so on. $x$-axis sensitivity of DH311 sensor is 0.97 pC/(m·s$^{-2}$), the range is 0~5000 m·s$^{-2}$ and the frequency response is 0~5 kHz. In the $x$-axis feed experiment, the spindle of NC lathe is stationary, the feed displacement is 25 mm, the feed speed is 300 mm/min, and the acceleration is 0.5 g. The axial acceleration signal of the workbench is collected, intercepted in condition of uniform motion and analyzed in the frequency domain.

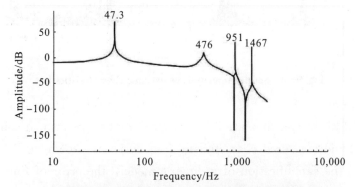

Fig. 5−2  Simulated result of frequency response

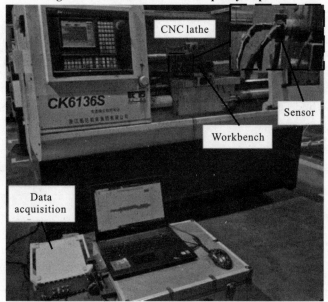

Fig. 5−3  Testing principle of the experiment table

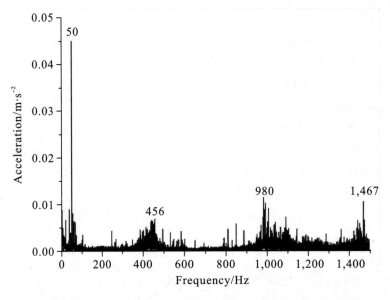

Fig. 5−4  Spectrum analysis of dynamic characteristics test

From Fig. 5−2 and Fig. 5−4, there are four characteristic frequencies within 0~2 kHz, and they are basically consistent. The comparison between theoretical and measured characteristic frequencies is shown in Table 5−2. Because of the simplification of geometric model, the error of finite element calculation and test, there are differences between the theoretical characteristic frequency and the experimental characteristic frequency. However, the corresponding frequency errors are less than 6%. Compared with reference[15], although Lagrange equation is used to establish the dynamic model of feed system, the error is more than 6.3%. It can be seen that the modeling method can accurately reflect the dynamic characteristics of the ball screw feed system.

Table 5−2  Comparison of characteristic frequency

| Order $i$ | Simulation results/Hz | Experimental results/Hz | Error $\sigma$/% |
| --- | --- | --- | --- |
| 1 | 47.3 | 50 | 5.7 |
| 2 | 476 | 456 | 4.2 |
| 3 | 951 | 980 | 3.0 |
| 4 | 1,419 | 1,467 | 3.3 |

## 5.4 Vibration characteristic analysis

The parameters are adjusted by integral multiples based on dynamic model as for the torsional vibration around the $x$ direction and axial vibration along the $x$ direction of the workbench has a different influence on machining accuracy. Meanwhile, the influence of the characteristic parameters on the acceleration range of the coupled torsional and axial vibration of the workbench is analyzed. In order to make the conclusion more universal, the product specifications and the calculation method of relevant literature[16-17] are consulted, and the initial parameters of the feed system commonly used in the numerical control machine tools for analysis are determined, as shown in Table 5-3.

Table 5-3  Parameter values used in the model

| Parameter | Value | Unit |
|---|---|---|
| $k_1$ | 416,666 | N/rad |
| $k_2$, $k_5$ | 50,000 | N/m |
| $k_3$, $k_6$ | 10 | rad/m |
| $k_8$ | $6.08 \times 10^8$ | N/m |
| $k_9$ | 385 | N/rad |
| $k_{11}$, $k_{12}$ | $4.56 \times 10^9$ | N/m |
| $b$ | 0.25 | m |

When the workbench position $x$ is fixed, the workbench is excited by a force. When the stiffness is increased from the original value to 5 times of the original value, the acceleration range of the torsional vibration of the workbench is shown in Table 5-4. As the other stiffness increases, it is equal to the acceleration rang of the axial vibration of the workbench (Table 5-5). In Tables 5-4 and 5-5, the rate of decline indicates the degree of decrease in the acceleration range of the vibration of the workbench when the parameter is increased from the original value to 5 times of the original value.

Table 5—4　The influence of the stiffness on the acceleration range of the torsional vibration of the workbench

| Acceleration range ($\times 10^{-13}$)/rad·s$^{-2}$ | Original | 2 times | 3 times | 4 times | 5 times | Rate of decline/% |
|---|---|---|---|---|---|---|
| $k_9$ | 7.1167 | 5.7514 | 4.8177 | 4.1394 | 3.6240 | 49.07 |
| $k_{11}$, $k_{12}$ | 7.1167 | 3.5592 | 2.3729 | 1.7798 | 1.4239 | 80.00 |
| $k_3$, $k_6$ | 7.1167 | 7.0917 | 7.0670 | 7.0427 | 7.0187 | 1.38 |
| $k_1$ | 7.1167 | 6.4871 | 5.9598 | 5.5118 | 5.1263 | 27.97 |

Table 5—5　The influence of the stiffness on the acceleration range of the axial vibration of the workbench

| Acceleration range ($\times 10^{-8}$)/rad·s$^{-2}$ | Original | 2 times | 3 times | 4 times | 5 times | Rate of decline/% |
|---|---|---|---|---|---|---|
| $k_8$ | 9.4145 | 4.7158 | 3.1458 | 2.3600 | 1.8884 | 79.94 |
| $k_9$ | 9.4145 | 9.4076 | 9.4007 | 9.3938 | 9.3870 | 2.93 |
| $k_3$, $k_6$ | 9.4145 | 9.4123 | 9.4103 | 9.4084 | 9.4066 | 0.08 |
| $k_2$, $k_5$ | 9.4145 | 9.3970 | 9.3864 | 9.3804 | 9.3775 | 0.04 |

As shown in Table 5—4, with the increase in the torsional stiffness of the screw-bearing joint, coupling, screw-nut joint and the radial stiffness of the workbench-guideway joint, the torsional vibration acceleration of the workbench decreases. In addition, the rate of descent in Table 5—4 indicates that the radial stiffness of the workbench-guideway joint and the torsional stiffness of the screw-nut joint are the main factors affecting the torsional vibration of the workbench. However, the torsional stiffness of the bearing affects is least. The reason is that the radial stiffness of the workbench-guideway joint and the torsional stiffness of the screw-nut joint are quite large, and the torsional stiffness of the nut in system is strongly coupled. For the torsional stiffness of the bearing, it is less than other torsional stiffness and the coupling of the bearing is smaller than the nut. The coupling has a large effect on the torsional vibration of the workbench, which is relatively small compared with the radial stiffness of the workbench-guideway joint and the torsional stiffness of screw-nut joint. This is because the coupling is a direct component of the transmitted torsional motion.

As shown in Table 5−5, with the increase of the axial and torsional stiffness of the screw-nut joint, the torsional stiffness of the bearing joint and the axial stiffness of the bearing joint, the axial vibration amplitude of the table decreases gradually.

It can be seen from the decreasing rate of Table 5−5 that among the factors affecting the axial vibration of the workbench, the axial stiffness of the screw-nut joint has the greatest influence on the axial vibration of the workbench, and the descending rate is also largest (79.94%), which is due to the strong coupling of the axial stiffness of the nut in the system and the direct connection with the worktable. The torsional stiffness of the screw-bearing joint has the least effect on the axial vibration of the workbench for three reasons. First, the value of torsional stiffness of the screw-bearing joint is very small. Second, its coupling is relatively small compared with the nut. Finally, the coupling is not directly connected to the workbench. Compared with the descent rates in Table 5−4 and Table 5−5, we can find that, for the same joint, the influence of the torsional stiffness on the axial vibration of workbench is significantly less than the axial stiffness of the axial vibration, such as the torsional stiffness and the axial stiffness of the screw-bearing joint and the screw-nut joint.

When the mass is reduced to a certain ratio, the acceleration range of the torsional vibration and the axial vibration of the workbench are shown in Table 5−6.

Table 5−6  The influence of the workbench mass on the acceleration range of the torsional vibration and the axial vibration of the workbench

| Mass | Original | $\frac{1}{2}$ times | $\frac{1}{3}$ times | $\frac{1}{4}$ times | $\frac{1}{5}$ times | Rate of decline/% |
|---|---|---|---|---|---|---|
| Torsional ($\times 10^{-13}$)/rad·s$^{-2}$ | 7.1167 | 6.9833 | 6.8719 | 6.7712 | 6.6869 | 6.09 |
| Axial ($\times 10^{-8}$)/m·s$^{-2}$ | 9.4145 | 9.2437 | 9.1063 | 8.9873 | 8.8790 | 5.69 |

Table 5−6 shows that the torsional vibration and the axial vibration of the workbench are reduced with the decrease of the mass of the workbench. The reason for this is that, the decrease of the mass of the workbench makes the moment of inertia reduced, as a result, the inertia force is reduced. Therefore, the vibration decreases.

When the length of the screw increases from 1 m to 2 m, the effect of axial stiffness of the nut on the acceleration range of the axial vibration of the workbench is shown in Table 5-7.

Table 5-7 The influence of the axial stiffness of the nut on the acceleration range of the axial vibration of the workbench under different screw's length

| Acceleration range ($\times 10^{-8}$)/m·s$^{-2}$ | Original | 2 times | 3 times | 4 times | 5 times | Rate of decline/% |
|---|---|---|---|---|---|---|
| Length (1 m) | 9.4145 | 4.7158 | 3.1458 | 2.3600 | 1.8884 | 79.9435 |
| Length (2 m) | 9.4464 | 4.7308 | 3.1555 | 2.3673 | 1.8942 | 79.9416 |

As shown in Table 5-7, as the length of the screw increases, the axial stiffness of the nut will gradually decrease the axial vibration of the workbench. At the same time, the influence of the length of the screw on the axial vibration of the workbench is less than the axial stiffness of the nut thorough analysis.

The influence of the torsional stiffness of the screw-nut joint on the torsional vibration of the workbench under variable lead is shown in Table 5-8. In Table 5-8, rate of decline represents that under variable lead, when the torsional stiffness of the nut is increased by 5 times from the original value, the degree of the torsional vibration is reduced by the quantitative acceleration of the workbench. The increase rate indicates the degree to which the torsional amplitude of the workbench increases as the lead increases from 5 mm to 10 mm under the variable torsional stiffness of the nut.

Table 5-8 The influence of the torsional stiffness of the nut on the acceleration range of the torsional vibration of the workbench under different screw's lead

| Acceleration ($\times 10^{-13}$)/rad·s$^{-2}$ | Original | 2 times | 3 times | 4 times | 5 times | Rate of decline/% |
|---|---|---|---|---|---|---|
| Max (lead 5 mm) | 3.5589 | 2.8763 | 2.4094 | 2.0699 | 1.8120 | 49.09 |
| Min (lead 5 mm) | -3.5578 | -2.8751 | -2.4083 | -2.0695 | -1.8120 | 49.07 |
| Range (lead 5 mm) | 7.1167 | 5.7514 | 4.8177 | 4.1394 | 3.6240 | 49.08 |

(to be continued)

(Continued)

| Acceleration $(\times 10^{-13})/\mathrm{rad} \cdot \mathrm{s}^{-2}$ | Original | 2 times | 3 times | 4 times | 5 times | Rate of decline/% |
|---|---|---|---|---|---|---|
| Max (lead 6 mm) | 3.5643 | 2.8841 | 2.4186 | 2.0801 | 1.8227 | 48.86 |
| Min (lead 6 mm) | −3.5640 | −2.8830 | −2.4180 | −2.0790 | −1.8220 | 48.87 |
| Range (lead 6 mm) | 7.1278 | 5.7675 | 4.8364 | 4.1591 | 3.6448 | 48.87 |
| Max (lead 8 mm) | 3.5708 | 2.8938 | 2.4301 | 2.0926 | 1.8360 | 48.58 |
| Min (lead 8 mm) | −3.5704 | −2.8937 | −2.4300 | −2.0924 | −1.8356 | 48.59 |
| Range (lead 8 mm) | 7.1412 | 5.7875 | 4.8601 | 4.1850 | 3.6716 | 48.59 |
| Max (lead 10 mm) | 3.5746 | 2.8995 | 2.4369 | 2.1001 | 1.8439 | 48.42 |
| Min (lead 10 mm) | −3.5745 | −2.8998 | −2.4372 | −2.1003 | −1.8440 | 48.41 |
| Range (lead 10 mm) | 7.1491 | 5.7993 | 4.8741 | 4.2004 | 3.6879 | 48.42 |
| Increase rate/% | 0.455 | 0.832 | 1.157 | 1.473 | 1.762 | — |

Table 5−8 shows that after increasing the lead, the torsional stiffness of the nut will have a reduced effect on the torsional vibration of the workbench, and conversely, after increasing the torsional stiffness of the nut, the influence of the lead on the torsional vibration of the workbench will increase.

When considering that the workbench position $x$ is not fixed, take $\theta = \omega t$, where $\omega$ is the motor speed. When the motor speed increases from 10 rad/s to 50 rad/s, the effect of motor speed on the acceleration range of the axial vibration of the workbench is shown in Fig. 5−5. It shows that with the increase of the motor speed, the axial vibration of the workbench will gradually increase.

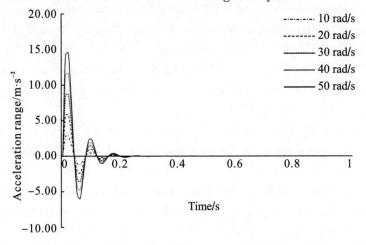

Fig. 5−5  **The influence of the motor speed on the axial vibration of the workbench**

# References

[1] Wang W, Zhou Y X, Wang H, et al. Vibration analysis of a coupled feed system with non-linear kinematic joints [J]. Mechanism and Machine Theory, 2019 (134): 562-581.

[2] Ma L, Cherubini G, Pantazi A, et al. Servo-patter design and track-following control for nanometer head positioning on flexible tape media [J]. IEEE Transactions on Control System Technology, 2012, 20 (2): 369-381.

[3] Cheng Q, Qi B B, Liu Z F, et al. An accuracy degradation analysis of ball screw mechanism considering time-varying motion and loading working conditions [J]. Mechanism and Machine Theory, 2019 (134): 1-23.

[4] Zhang H D, Zhou H L. The vibration frequency analysis of the screw feed system with spring supports [J]. Journal of Vibroengineering, 2017, 19 (3): 1479-1493.

[5] Msukwa M R, Uchiyama N. Design and experimental verification of adaptive sliding mode control for precision motion and energy saving in feed drive systems [J]. IEEE Access, 2019 (7): 20178-20186.

[6] Liu L L, Liu H Z, Wu Z Y, et al. Research on stability of feed servo system of machine tool at low speed [J]. Journal of Vibration and Shock, 2010, 29 (5): 187-190.

[7] Ansoategui I, Campa F J. Mechatronics of a ball screw drive using an N degrees of freedom dynamic model [J]. International Journal of Advanced Manufacturing Technology, 2017, 93 (1-4): 1307-1318.

[8] Zou C F, Zhang H J, Lu D, et al. Effect of the screw-nut joint stiffness on the position-dependent dynamics of a vertical ball screw feed system without counterweight [J]. Proceedings of the Institution of Mechanical Engineers Part C — Journal of Mechanical Engineering Science, 2018, 232 (15): 2599-2609.

[9] Vicente D A, Hecker R L, Villegas F J, et al. Modeling and vibration mode analysis of a ball screw drive [J]. The International Journal of Advanced Manufacturing Technology, 2012, 58 (1-4): 257-265.

[10] Weng K K, Cheng Y, Xia L L, et al. The dynamic characteristics analysis of feed system of a vertical machining center based on conjoint interfaces [J]. Machinery Design & Manufacture, 2012 (3): 130-132.

[11] Jang W Y, Park M Y, Kim J H, et al. Analysis of vibration characteristics of one-axis heavy duty stages [J]. Journal of Mechanical Science and Technology, 2017, 31 (12): 5721-5727.

[12] Zhang H R. Dynamic analysis of machine tool feed system [D]. Changchun: Jilin

University, 2009.

[13] Yang X J, Zhao W H, Liu H. Coupled vibration mode analysis of a ball screw drive system [J]. Machinery Design & Manufacture, 2012 (10): 259-261.

[14] Zhu J M, Zhang T C, Li X R. Dynamic characteristic analysis of ball screw feed system based on stiffness characteristic of mechanical joints [J]. Journal of Mechanical Engineering, 2015, 51 (17): 72-82.

[15] Lu H, Fan W, Zhang X B, et al. Dynamic characteristics analysis and test of dual-driving feed system driven by center of gravity [J]. Mathematical Problems in Engineering, 2018: 1-16.

[16] Gong Z C, Niu W T, Li J J, et al. Dynamic characteristics simulation and performance optimization of the feed system based on the electromechanical coupling [J]. Journal of Machine Design, 2018, 35 (9): 8-16.

[17] Wang M, Le B B, Pei E Y. Contact stiffness modeling and analysis of linear ball guides based on Hertz contact theory [J]. Journal of Beijing University of Technology, 2015, 41 (8): 1128-1132.

# Chapter 6　Influence of Damping Property on Chatter of Feed System

## 6.1　Introduction

Damping has an important influence on the stability and flutter suppression of feed system[1]. The damping of the feed system can be divided into joint damping and component damping[2-3]. The research shows that more than 60% of the vibrations of machine tools are caused by the stiffness and damping of the joint, and more than 90% of the total damping value is caused by the damping of the joint[4-5]. In feed system of CNC machine tools, the joint surface can be divided into fixed and movable joint surfaces, among which, the fixed joint surface mainly has tool holder-blade joint, the movable joint includes the screw-nut and slide-guide rail[6-7]. At present, the parameter identification of fixed interface is the research focus, and the parameter identification of movable interface is still in the initial stage[8]. Yang et al. [9] established a multi-damper to suppress the objective function of the main structure and optimized the stiffness and damping of the multiple dampers by constructing a mini-max method, which improved the ability of multiple dampers to suppress the flutter of the machine. This ensures the quality of the process but does not study the optimal position of the multiple dampers. Haranath et al. [10] proposed laying a layer of viscous material on the surface of the tool to suppress tremors, but the materials used in the study were much more expensive than other materials. Hong et al. [11] discussed the stabilizing effect of the time-varying spindle speed through experiments. It is concluded that the jitter reduction amplitude value is related to the change of the spindle speed mode, but the bandwidth has certain limitations on the method. Therefore, it is of great theoretical value to study the effect of damping on the vibration of the feed system in order to optimize the system structure in the future.

## 6.2　Dynamic model based on the Lagrange equation

The feed system adopts a direct connection structure, and the servo motor is connected with the rigid coupling and the screw is driven by the coupling to perform a rotary motion. The screw nut mechanism drives the table to make a linear motion. Fig. 6−1 shows the structure of the feed system. The servo motor 1 (ACSM130−G06025LZ) is fixed at one end of the machine and its rated speed is 2,500 rad/min. The ball screw adopts an inner circulation single nut type structure with a lead $l$ of 5 mm, a nominal diameter $d_0$ of 20 mm, a ball diameter $d$ of 3.175 mm, and the outer diameter of the screw is 19.5 mm. The lead screw 4 is supported by the support bearing 3 by a fixed end and a free end mounting manner, and the span is 800 mm. The servo motor is coupled to the coupling 2 and driven by the coupling to perform a rotary motion. The worktable 6 is linearly moved in the $X$-direction by the nut 5. The mass of the worktable is 20 kg.

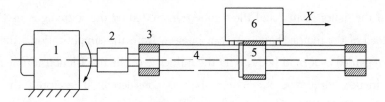

**Fig. 6−1　Schematic diagram of the feed system**

In order to facilitate the analysis, the feed system is simplified and the following assumptions are made: the return clearance is not considered; the total stiffness of the feed system remains unchanged[12]; the coupling is considered as a torsion spring; the stiffness of the motor shaft is large and it is considered as a rigid body. Considering the effect of nut pre-tightening and bearing friction, the friction torque generated by the nut pre-tightening and the friction torque generated by the bearing is converted to the lead screw for consideration. Considering the influence of the rotary deformation, the torsion damping and axial damping of the screw are converted into the comprehensive axial damping of the screw. The torsion stiffness of the lead screw and the axial stiffness of the feed are converted into the feed comprehensive axial stiffness and connected in series on the workbench. So the lead screw is considered as a rigid body. A simplified mechanical model is shown in Fig. 6−2.

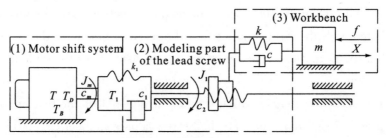

Fig. 6−2 Simplified mechanical model of the feed system

Each motion part of the feed system has a certain influence on the output response. In order to quantitatively analyze the dynamic characteristics of the system and reveal the influence of various factors of dynamic characteristics on the system flutter, a multi-degree-of-freedom system dynamics model is established based on Newton's second law.

(1) Motor shaft modeling. The rated torque of the motor is used as the input torque of the motor shaft, and the torque transmitted by the coupling is used as the output load of the motor shaft. As can be seen from the simplification, the motor shaft is regarded as a rigid body, and the influence of the nut friction torque and the bearing friction torque on the motor shaft is considered. Therefore, the torque balance equations of the motor shaft can be obtained:

$$\left.\begin{array}{l} T - T_D - T_B = J_m \dfrac{d^2 \varphi_1}{dt^2} + c_m \dfrac{d\varphi_1}{dt} + T_1 \\ T_D = \dfrac{1-\eta^2}{\eta^2} \cdot F_p \cdot \gamma \\ T_B = a \cdot P \cdot \mu \cdot d \end{array}\right\} \quad (6-1)$$

where, $T$ is the rated input torque of the motor; $T_D$ is the torque of the friction torque generated by the pre-tightening force converted to the motor shaft; $T_B$ is the torque of the combined friction torque generated by the supporting bearing to the motor shaft; $T_1$ is the torque transmitted by the coupling; $\varphi_1(t)$ is the rotation angle of the input shaft; $J_m$ is the equivalent moment of inertia of the motor shaft; $c_m$ is the torsional viscous damping coefficient of the motor shaft; $F_p$ is the pre-tightening force of the nut; $\gamma$ is the conversion coefficient of the rotational to linear motion, $\gamma = \dfrac{l}{2\pi}$; $\eta$ is the transmission efficiency; $a$ is the conversion coefficient of the friction torque to the motor shaft; $P$ is the equivalent dynamic load of the bearing; $\mu$ is the rolling friction coefficient; $d$ is the diameter of motor shaft.

(2) Modeling of screw drive. It is known from the simplifying assumption that the screw is transformed from a flexible body to a rigid body. Therefore, the torsional torque transmitted by the coupling during modeling is used as the input torque for driving the screw rotation. The driving force for driving the movement of the table is used as the output load of the screw drive, and the dynamic model is obtained:

$$\left. \begin{aligned} T_1 &= J_1 \frac{d^2 \varphi_2}{dt^2} + c_2 \frac{d\varphi_2}{dt} + \gamma F \\ T_1 &= c_1 \frac{d\varphi}{dt} + k_1 \varphi(t) \\ F &= k[\gamma \varphi_2(t) - x(t)] + c \frac{d[\gamma \varphi_2(t) - x(t)]}{dt} \end{aligned} \right\} \quad (6-2)$$

where, $J_1$ is the equivalent moment of inertia of the screw; $\varphi$ is the relative torsion angle of the motor shaft and the lead screw, $\varphi = \varphi_1 - \varphi_2$; $k_1$ is the torsional stiffness of the coupling; $c_2$ is the torsional viscous damping coefficient of the screw; $c_1$ is the torsional damping coefficient of the coupling; $c$ is the comprehensive damping coefficient of the screw; $k$ is the comprehensive axial stiffness of the feed system; $\delta$ is the axial deformation of the linear motion of the rotation, $\delta(t) = \gamma \varphi_2(t) - x(t)$; $F$ is the driving force of the workbench.

(3) Worktable modeling. During the feeding process, there is viscous damping and friction between the slider and the guide rail due to lubrication and contact. As in Fig. 6-2, from the simplified force relationship of the workbench, the dynamic equilibrium of the table can be obtained in the

following equation:

$$F - c_t \frac{dx}{dt} - f = m \frac{d^2x}{dt^2} \qquad (6-3)$$

where, $c_t$ is the viscous damping coefficient of the contact surface; $f$ is the frictional force of the slider, $f = \mu_1 mg$; $m$ is the mass of the workbench; $x(t)$ is the displacement of the table.

## 6.3 Simulation analysis

### 6.3.1 Simulation parameter setting

Some dynamic characteristic parameter values of the mechanical system can't be accurately calculated by using mathematical formulas. Therefore, the simulation parameters are estimated and processed, and the obtained system simulation parameters are shown in Table 6−1.

Table 6−1 System simulation parameters

| Parameter | Value |
| --- | --- |
| Motor moment of inertia $J_m$/kg·m² | 0.00003 |
| Screw spindle inertia $J_1$/kg·m² | 0.00012 |
| Coupling torsional stiffness $k_1$/N·m·rad⁻¹ | 416,666 |
| Feed system integrated axial stiffness $k$/N·m⁻¹ | 88,495,575 |
| Screw torsion viscous damping $c_2$/N·s·rad⁻¹ | 0.007 |
| Guide rail viscous damping $c_t$/N·s·m⁻¹ | 3,030 |
| Coupling torsional damping $c_1$/N·s·m⁻¹ | 0.012 |
| Screw integrated axial damping $c$/N·s·m⁻¹ | 10,000 |

(1) Determination of the moment of inertia parameters. The correct description of the moment of inertia of a rigid body is one of the most pivotal

aspects of studying the motion of a rigid body. In order to obtain a reasonable value of the moment of inertia, the screw and the motor shaft are simplified to a uniform rod for calculation[13].

(2) Determination of system stiffness parameters. When estimating the torsional stiffness of the coupling, the formula in material mechanics is used for calculation. For ensuring the screw feed stiffness, the table is moved to the maximum stroke for calculation. When calculating the nut stiffness, the preload is calculated as 10% of the basic dynamic load rating. Feed system integrated axial stiffness is calculated from equation (6−4)[14].

$$\frac{1}{k} = \frac{1}{k_{x1}} + \frac{1}{k_{x2}} + \frac{1}{k_{x3}} + \frac{y^2}{k_\varphi} \qquad (6-4)$$

where, $k_{x1}$ is the contact stiffness of the nut, the value of $k_{x1}$ is $5.6 \times 10^8$ N/m; $k_{x2}$ is the contact stiffness of the bearing combination; $k_{x3}$ is the axial stiffness of the screw, the value of $k_{x3}$ is $1.57 \times 10^8$ N/m; $k_\varphi$ is the torsional stiffness of the screw, and the value of $k_\varphi$ is 1945.8 N·m/rad.

(3) Determination of damping. The modal simulation experiment of the lead screw with ANSYS can obtain the maximum damping ratio $\xi = 0.0958$ of the lead screw. The screw damping is calculated according to the mechanical vibration theory. When adding the guide rail damping, it is considered from the low viscosity oil. So the guide rail viscous damping value can be consulted according to the lubrication conditions[15].

### 6.3.2 Simulation result analysis

MATLAB/SIMULINK is used as a research tool, and the simulation model is shown in Fig. 6−3. Set the parameters of the simulation and input the analog motor torque with the step signal. The simulation time is 0.5 s.

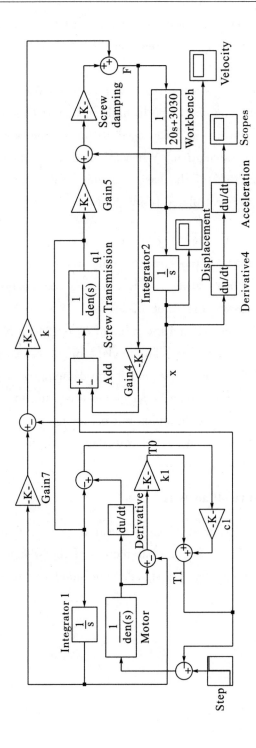

**Fig.6-3 SIMULINK simulation model**

(1) According to the analysis of the actual design system and the calculation of the initial parameters of the dynamic simulation, the simulation curves of initial velocity and acceleration are obtained. In order to reflect the true characteristics of the initial velocity and acceleration simulation curves, they are differently amplified: The acceleration interval corresponding to the y-axis is unchanged, and the time interval corresponding to the x-axis is enlarged in the x-axis, as shown in Fig. 6−4. Similarly, the initial velocity curve is amplified in the y-axis, as shown in Fig. 6−5. In addition, aiming to clearly reflect the dynamic characteristics of the initial acceleration curve, a specific simulation curve on the initial acceleration curve is enlarged in the x-axis and the y-axis, as shown in Fig. 6−6.

As can be seen from Fig. 6 − 5, the workbench will produce severe chattering for a period of time after the quick start. It can be obtained from Fig. 6−4 and Fig. 6−6 that the maximum amplitude of the table acceleration is relatively stable.

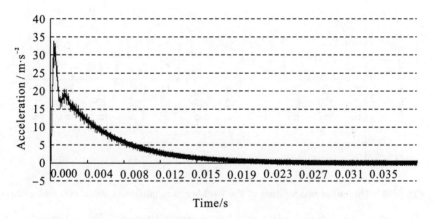

Fig. 6−4　The initial acceleration of the workbench amplified in the x-axis

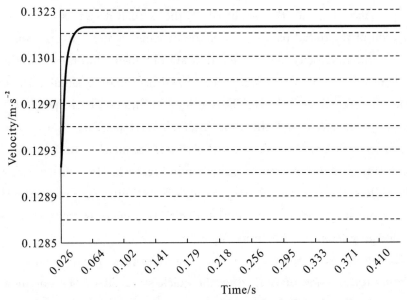

Fig. 6-5  The initial velocity of the workbench amplified in the y-axis

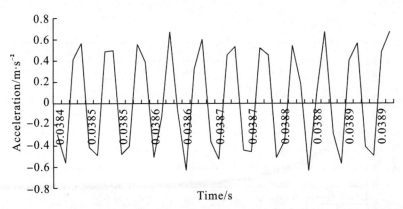

Fig. 6-6  The initial acceleration of the workbench amplified in the x-axis and y-axis

From Fig. 6-5, the table can be severely fluttered for a period of time after a quick start. As can be seen from Fig. 6-4 and Fig. 6-6, the maximum amplitude of the table acceleration is relatively stable.

(2) In order to further quantitatively compare and analyze the influence of the dynamic characteristic parameters of the feed system on the flutter, the interval corresponding to the enlarged simulation curve of Fig. 6-6 is taken as the simulation analysis interval. When other parameters are unchanged, the quality of the system table, the torsion damping coefficient of the coupling,

the integrated damping coefficient of the lead screw, the viscous damping coefficient of the joint surface of the guide rail, and the viscous damping of the screw increas by 5%, and then simulate. The simulation results of the acceleration flutter are obtained by the factors in the same corresponding interval as in Fig. 6-6, and the factors affecting the flutter of the workbench (the integrated screw damping, the coupling damping, and the screw torsional viscous damping) are taken from the simulation results when 5% is increased to 20%, as shown in from Fig. 6-7 to Fig. 6-9. In addition, the influence of each factor on the acceleration flutter of the table is obtained by taking the maximum acceleration amplitude obtained by different influencing factors in the same interval, as shown in Fig. 6-10.

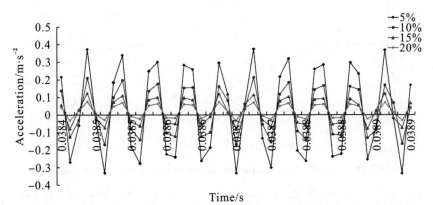

Fig. 6-7 Increasing the comprehensive damping of the lead screw in equal proportion

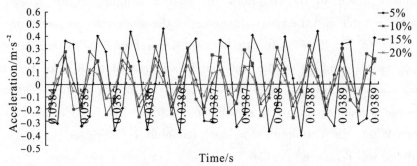

Fig. 6-8 Increasing coupling damping in equal proportion

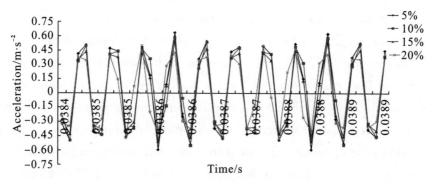

Fig. 6−9  Increasing torsional viscous damping of the lead screw in equal proportion

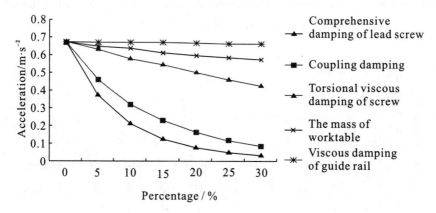

Fig. 6−10  Effects of various factors on acceleration flutter of the worktable

It can be seen from the simulation results that the table flutter has been reduced by increasing the comprehensive damping of the lead screw, the torsional damping of the coupling, the viscous damping of the screw, the mass of the table and the viscous damping of the guide rail, as shown in from Fig. 6 − 7 to Fig. 6 − 9. It can be seen from Fig. 6 − 10 that after the comprehensive damping of the lead screw is increased to 30% of the initial value, the maximum amplitude of the acceleration flutter of the worktable in the range is reduced rapidly from 0.67 to 0.037, which is about 18.11 times lower. When the coupling's torsional damping is increased by 30%, its amplitude will decrease to 0.086, which is about 7.79 times lower. When the rotational viscous damping is increased, its amplitude decreases by 1.56. When the mass of the worktable and the contact viscous damping of the guide rail increase, their values decrease by about 1.03 times and 1.01 times respectively, and the tendency of chatter attenuation is very gentle. Therefore, the influence of the change of lead screw damping and coupling torsional damping on the chatter is far greater than that of the worktable mass and the viscous damping of the guide rail, and the lead screw damping has the most obvious influence on the chatter of the worktable.

# References

[1] Liang Q M, Yin H J. Modeling and simulation of the feed drive system for a type of CNC engraving and milling machine [J]. Mechanical Research & Application, 2013, 26 (2): 45-47.

[2] Chen D X, Liu S, Chen L B, et al. Impact analysis of characteristics of CNC machine tool on machining quality [J]. Aeronautical Manufacturing Technology, 2014, 5 (3): 63-66.

[3] Xia J Y, Wu B, Hu Y M, et al. Experimental research on factors influencing thermal dynamics characteristics of feed system [J]. Optical Precision Engineering, 2010, 34 (2): 357-368.

[4] Weng D K, Cheng Y, Xia L L, et al. The dynamic characteristics analysis of feed system of a vertical machining center based on conjoint interfaces [J]. Machinery Design & Manufacture, 2012 (3): 130-132.

[5] Ding X H, Yuan J T, Wang Z H, et al. Analysis of static and dynamic characteristics of linear feed system with double ball screw [J]. Machinery Design & Manufacture, 2014 (3): 155-157.

[6] Zhao Z M, Wang M F. Application of CNC cutting chatter monitoring and control technology [J]. Tool Engineering, 2012 (4): 84-86.

[7] Wang Y H, Wang M. Advances on machining chatter suppression research [J]. Chinese Journal of Mechanical Engineering, 2010, 46 (7): 167-173.

[8] Moradoi H, Gholamreza V, Mrhfi B. Vibration absorption design to suppress regenerative chatter in non-linear milling process: Application for machining of cantilever plates [J]. Applied Mathematical Modeling, 2015, 39 (2): 600-620.

[9] Yang Y Q, Liu Q, Wang M. Optimization of multiple tuned mass dampers for chatter suppression in turning [J]. Journal of Vibration Engineering, 2010, 23 (4): 468-474.

[10] Haranath S, Ganesan N, Rao B V A. Dynamic analysis of machine tool structures with applied damping treatment [J]. International Journal of Machine Tools and

Manufacture, 1987, 27 (1): 43-55.

[11] Hong L, Yu J L, Gao L B. Stability effect of on-line time-varying spindle speed [J]. Mechanical Engineer, 1994 (5): 38-40.

[12] Wang Y Q, Zhang C R. Simulation modeling of a ball screw feed drive system [J]. Journal of Vibration and Shock, 2013, 32 (3): 46-55.

[13] Fan Q, Ren Y J. Study of calculation method for rotational inertia of a uniform rod to random rotation shaft [J]. Journal of Chongqing Technology and Business University (Natural Science Edition), 2007, 24 (5): 525-528.

[14] Wu Z Y, Liu H Z, Liu L L. Modeling and analysis of cross feed servo system of heavy duty lathe subjected to friction [J]. Chinese Journal of Mechanical Engineering, 2012, 48 (7): 87-93.

[15] Satpute N V, Satpute S N. Design and analysis of ball screw-based inertial harvester [J]. Iranian Journal of Science and Technology-Transactions of Mechanical Engineering, 2019, 43 (2): 359-374.

# Chapter 7　Axial Non-linear Vibration of Feed System

## 7.1　Introduction

According to the traditional flutter research, the rigidity of machine structure is linear, and the flutter in the process of servo drive is also linear[1]. However, the linear flutter theory can not give a reasonable explanation for some phenomena in the actual feeding and cutting process. For example, according to the linear flutter theory, once the cutting depth exceeds a certain critical value, the system will become unstable, and the amplitude will increase infinitely. However, the amplitude of the flutter phenomenon encountered in practice is stable at a certain level. In view of the above phenomena, many researchers believe that there are non-linear factors in machine chatter. In the feed system, there are a lot of non-linear factors. These non-linear factors include both non-linear stiffness and non-linear damping. As one of the core parts of the typical feed system, the lead screw is a kind of slender bar with obvious vibration. The stiffness of ball-screw has strong non-linearity. The 30%~50% total stiffness of machine tool feed system belongs to ball-screw stiffness. The ball-screw stiffness is one of the important factors influencing the dynamic characteristics of the system[2-3]. If the mechanical model of machine tool feed system is simplified to a linear system and the non-linear factors are ignored, the research results and the actual situation will have serious errors and the movement features of machine tool feed system can not be fully and accurately known. Research on the non-linear characteristics of ball-screw of machine tool feed system will contribute to the improvement of the performance of the machine tool vibration control and processing.

In recent years, a more accurate modeling method is to put the screw as Timoshenko beam and consider the shear deformation and moment of inertia of the ball-screw. Leonard[4-6] studied the ball-screw bearing support and its pre-tightening force of impact on the stiffness of the ball-screw. Yang[7] studied the stiffness calculation method of feed system ball-screw. Wu et al. [8] pointed out that the stiffness of feed system ball-screw affects the positioning accuracy of the system. Esmailzaden and Ohadt[9] studied the vibration of Timoshenko beam under the action of axial force. Arboleda et al. [10] studied the stability and inherent frequency of Timoshenko beam under the action of axial force. Wang et al. [11-12] proved the ball-screw of machine tool feed system is characteristic with non-linear dynamics system through the theoretical analysis and experimental verification of NC machine. Thus, the ball screw of Timoshenko beam is simplified. It is necessary to consider the non-linear vibration characteristics of the ball screw under the shear deformation.

## 7.2 Theoretical modeling

The ball-screw of machine tool feed system is simplified to Timoshenko beam. The material of the beam is uniform and the section is a circular cross. The beam is fixed at one end and is supported at the other end. The length of ball-screw is $L$, the axial force is $F$, the gravity of the workbench is $G$ and the torque is $M$, as is shown in Fig. 7-1.

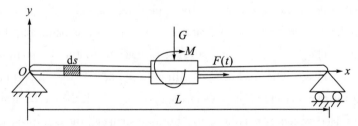

Fig. 7-1 The simplified model

Axial force $N$, shear forece $Q$ and bending moment $M$ are applied to the segment as at the beam. The ball-screw vibration displacement along the $x$-axis is set as $u(x, t)$, and the ball-screw vibration displacement along the $y$-axis is set as $v(x, t)$. Under the joint influence of bending moment $M$ and shear force $Q$, the ball-screw deformation angle is $\theta$, as is shown in Fig. 7-2.

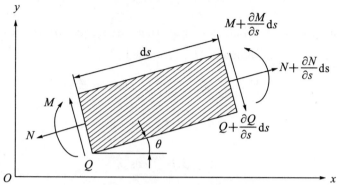

Fig. 7-2  Ball-screw segments force analysis

According to the micro force and Newton's second law, the following differential equation about the screw motion are established.

$$\rho A \frac{\partial^2 u}{\partial t^2} = \frac{\partial}{\partial x}(N\cos\theta + Q\sin\theta)\cos\theta \qquad (7-1)$$

$$\rho A \frac{\partial^2 v}{\partial t^2} = \frac{\partial}{\partial x}(N\sin\theta - Q\cos\theta)\cos\theta \qquad (7-2)$$

$$\rho J \frac{\partial^2 \psi}{\partial t^2} = \frac{\partial M}{\partial x}\cos\theta - Q \qquad (7-3)$$

Due to the small deformation of the ball screw, the trigonometric function is expanded by Taylor formula and the higher-order variables are omitted. Equation (7-4) and equation (7-5) are obtained.

$$\sin\theta \approx \frac{\partial v}{\partial x} \qquad (7-4)$$

$$\cos\theta \approx 1 - \frac{1}{2}\left(\frac{\partial v}{\partial x}\right)^2 \qquad (7-5)$$

Assuming that the ball-screw cross section and the axis lead always keep vertical, the axial displacement $\tilde{u}$ of any segment is composed of three parts:

(1) Axial displacement of cross section caused by axial force $N$ is $u$.

(2) Axial displacement caused by rotation of cross section is $y\sin\psi$.

(3) Axial displacement caused by transverse bending is $\int_0^x \sqrt{1+\left(\frac{\partial v}{\partial x}\right)^2}\,dx - x$.

Total axial displacement from one segment:

$$\tilde{u} = u + y\sin\psi + \int_0^x \sqrt{1+\left(\frac{\partial v}{\partial x}\right)^2}\,dx - x \qquad (7-6)$$

Normal strain of the segment:

$$\varepsilon = \frac{\partial \tilde{u}}{\partial s} \approx \left[\frac{\partial u}{\partial x} + y\cos\psi \frac{\partial \psi}{\partial x} + \frac{1}{2}\left(\frac{\partial v}{\partial x}\right)^2\right]\cos\theta \qquad (7-7)$$

According to the mechanics of materials, the equation of axial force $N$, shear force $Q$, bending moment $M$ is obtained.

$$N = EA\left[\frac{\partial u}{\partial x} + \frac{1}{2}\left(\frac{\partial v}{\partial x}\right)^2\right]\cos\theta \approx EA\frac{\partial u}{\partial x} \qquad (7-8)$$

$$Q = kAG\left(\psi - \frac{\partial v}{\partial x}\right) \qquad (7-9)$$

$$M = EJ\frac{\partial \psi}{\partial x}\cos\theta \qquad (7-10)$$

Substituting equation (7−4) ~ equation (7−10) into equation (7−1) ~ equation (7−3), the dynamic equation (7−11) ~ equation (7−13) are obtained.

$$\rho A \frac{\partial^2 u}{\partial t^2} = EA \frac{\partial^2 u}{\partial x^2}\left[1 - \frac{1}{2}\left(\frac{\partial v}{\partial x}\right)^2\right] + kAG\left(\frac{\partial \psi}{\partial x} - \frac{\partial^2 v}{\partial x^2}\right)\left[1 - \frac{1}{2}\left(\frac{\partial v}{\partial x}\right)^2\right]\frac{\partial v}{\partial x} \qquad (7-11)$$

$$\rho A \frac{\partial^2 v}{\partial t^2} = EA \frac{\partial^2 u}{\partial x^2} \cdot \frac{\partial v}{\partial x} - kAG\left(\frac{\partial \psi}{\partial x} - \frac{\partial^2 v}{\partial x^2}\right)\left[1 - \frac{1}{2}\left(\frac{\partial v}{\partial x}\right)^2\right]^2 \qquad (7-12)$$

$$\rho J \frac{\partial^2 \psi}{\partial t^2} = \left\{EJG\frac{\partial^2 \psi}{\partial x^2}\left[1 - \frac{1}{2}\left(\frac{\partial v}{\partial x}\right)^2\right] - EJG\frac{\partial \psi}{\partial x} \cdot \frac{\partial v}{\partial x} \cdot \frac{\partial^2 v}{\partial x^2}\right\}\left[1 - \frac{1}{2}\left(\frac{\partial v}{\partial x}\right)^2\right] - kAG\left(\frac{\partial \psi}{\partial x} - \frac{\partial^2 v}{\partial x^2}\right) \qquad (7-13)$$

$x$ and $t$ are variables of the function $v$. When studying the lateral vibration of a ball screw, the general solution is

$$v = \sin\frac{n\pi x}{L}\sin\omega_n t \qquad (7-14)$$

First-order natural vibration mode function:

$$v = \sin\left(\frac{\pi x}{L}\right) \cdot \left(\frac{\pi}{L}\right)^2 \sin\sqrt{\frac{EI}{\rho A}}t \qquad (7-15)$$

Due to the shear deformation:

$$\psi = \frac{\partial v}{\partial x} \qquad (7-16)$$

$$\psi = \frac{\cos\left(\frac{\pi x}{L}\right)\pi\sin\left(\pi^2\sqrt{\frac{EI}{\rho A}}\frac{t}{L^2}\right)}{L} \qquad (7-17)$$

To simplify equation (7−11), the influence of the five order term is ignored, and a ball-screw axial vibration equation without damping is obtained:

$$\rho A u'' + \frac{EA\pi^2}{L^2}u + \frac{EA\pi^4}{L^4}u^3 = 0 \qquad (7-18)$$

Considering the ball-screw axial vibration with damping force and exciting force is $F\sin\omega t$, the axial vibration equation satisfies the form of duffing equation:

$$u'' + cu' + \frac{E\pi^2}{\rho L^2}u + \frac{E\pi^4}{\rho L^4}u^3 = F\sin\omega t \qquad (7-19)$$

Equation (7-19) is the ball-screw axial non-linear vibration equation of machine tool feed system.

## 7.3 Numerical simulation

Graphic method is used to analyze the non-linear vibration of the ball-screw of feed system. The duffing equation is numerically simulated by MATLAB/SIMULINK (Fig. 7-3). The vibration equation is analyzed by phase track diagram and Poincare section. Simulation parameters are set up. Solver is ode113, with solution accuracy to $10^{-9}$. The initial value of displacement and velocity are 0. The equation variables include ball-screw length $L$, damping ratio $c$, the amplitude of exciting force $F$ and the frequency of exciting force $\omega$. The Poincare section is obtained by MATLAB code.

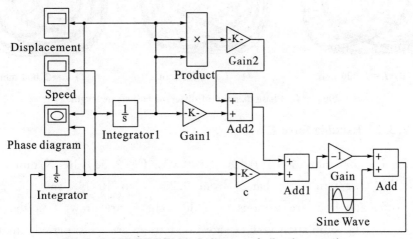

Fig. 7-3 SIMULINK block diagram of vibration equation

### 7.3.1 Ball-screw length $L$

Set $F=1$ N, $\omega=1$ rad/s, $c=0.02$, $E=2.06\times 10^5$ N/mm$^2$, $p=7.75\times 10^{-6}$ kg/mm$^3$. $L$ changed from 250 mm to 2000 mm. The phase diagram and Poincare sections with different $L$ are drawn, as shown in Fig. 7−4. The simulation results show that quasi-periodic chaotic motion appears at the axial vibration of feed system. The singularities are distributed on both sides at the initial points. After a period of time, motion reaches quasi-periodic state. With the increase of the length of the ball-screw, the stability of axial vibration became weaker.

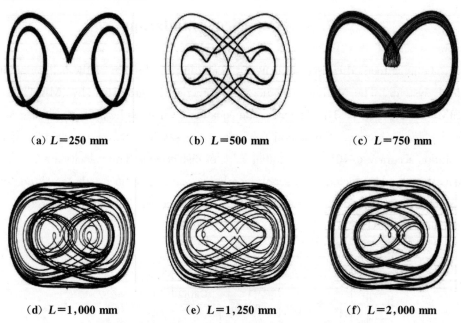

(a) $L=250$ mm  (b) $L=500$ mm  (c) $L=750$ mm

(d) $L=1,000$ mm  (e) $L=1,250$ mm  (f) $L=2,000$ mm

Fig. 7−4  Phase diagram of different ball-screw length

### 7.3.2 Exciting force $F$

Set $L=500$ mm, $\omega=1$ rad/s, $c=0.02$, $E=2.06\times 10^5$ N/mm$^2$, $p=7.85\times 10^{-6}$ kg/mm$^3$. $F$ changs from 0.25 N to 10 N, and the phase trajectories and Poincare sections with different $F$ are drawn, as shown in Fig. 7−5. The simulation results show that there are quasi-periodic motions and chaotic motions in the axial vibration of feed system. The singularities are distributed on both sides at the initial points. With the increase of exciting force $F$, the stability of axial vibration became weaker and Periodic movement is irregular.

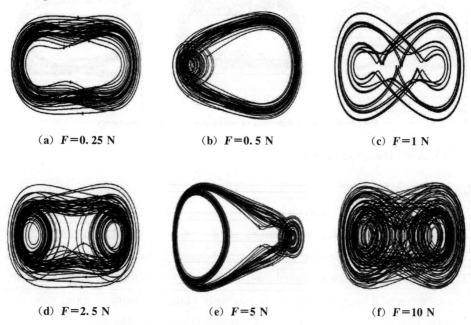

Fig. 7−5　Phase diagram of different exciting force

### 7.3.3　Excitation frequency $\omega$

Set $L=1000$ mm, $F=1$ N, $c=0.02$, $E=2.06\times10^5$ N/mm$^2$, $p=7.85\times10^{-6}$ kg/mm$^3$. $\omega$ changes from 0.25 rad/s to 2 rad/s. The phase diagrams and Poincare sections of different $\omega$ values are drawn. As seen in Fig. 7−6, the simulation results show that there are quasi periodic and chaotic motions in the axial vibration of ball screw. The singularities are distributed on both sides at the initial point. When the excitation frequency is 0.75 rad/s [Fig. 7−6(c)] or multiple of 0.75 rad/s [Fig. 7−6(e)], the stability of axial vibration becomes weak.

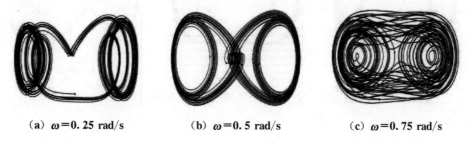

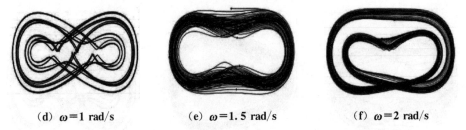

(d) $\omega=1$ rad/s  (e) $\omega=1.5$ rad/s  (f) $\omega=2$ rad/s

Fig. 7-6  Phase diagram of different excitation frequency

### 7.3.4 Damping ratio c

Set $L=300$ mm, $F=1$ N, $\omega=1$ rad/s, $E=2.06\times10^5$ N/mm², $p=7.85\times10^{-6}$ kg/mm³. $c$ changes from 0.005 to 0.1. The phase diagrams and Poincare sections of different $c$ values are drawn. The simulation results show that there are quasi periodic and chaotic motions in the axial vibration of ball screw. The singularity is distributed on both sides of the initial point. With the increase of damping ratio $c$, the stability and periodicity of axial vibration become better.

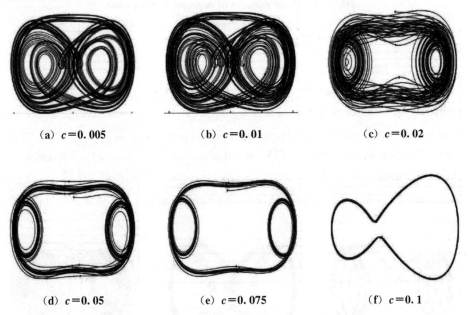

(a) $c=0.005$  (b) $c=0.01$  (c) $c=0.02$

(d) $c=0.05$  (e) $c=0.075$  (f) $c=0.1$

Fig. 7-7  Phase diagram of different damping ratio

## 7.4 Experimental verification

Exciting forces on multiple points and collecting the responding signals from a single point is adopted, and the feed system test-table of machine tool is taken as the research object. Excited with a hammer and tested for axial vibration by a piezoelectric acceleration sensor, the test environment is almost with no interference. Modal test system is shown in Fig. 7-8. Spectrogram of axial vibration is shown in Fig. 7-9. Spectrum gram of axial vibration is a continuous spectrum, which proves that the axial vibration of the ball-screw of the machine tool feed system has chaotic motion.

Fig. 7-8    Modal test system

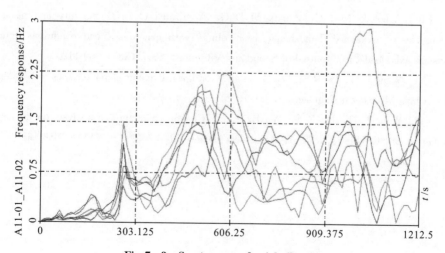

Fig. 7-9    Spectrogram of axial vibration

# References

[1] Huang Z Y. Development and application of precision high-speed ball screw [J]. Manufacturing Technology & Machine Tool, 2002 (5): 8-11.

[2] Pritschow G. A comparison of linear and conventional electromechanical drives [J]. Annals of the CIRP, 1998, 47 (2): 541-548.

[3] Liao B Y, Zhou X M, Yin Z H. Modern mechanical dynamics and its engineering application [M]. Beijing: Mechanical Industry Press, 2003.

[4] Leonard C P. Particularities of modeling ball screw based NC axes as finite degrees of freedom dynamic systems [J]. Buletinul Institutului Polotehnic Din Iasi, 2005 (5): 1-6.

[5] Leonard C P, Mircea C, Liviu M. Methods of evaluation of the mechanical characteristics influences on the NC balls screw drives dynamic behavior [J]. Buletinul Institutului Polotehnic Din Iasi, 2005 (5): 13-18.

[6] Leonard C P, Liviu M. Axial pre-stress of NC machine tools ball screw drives [J]. Buletinul Institutului Polotehnic Din Iasi, 2005 (5): 7-12.

[7] Yang Z X. Calculation of transmission rigidity of ball screw pair [J]. Manufacturing Technology & Machine Tool, 1999 (7): 12-14.

[8] Wu N X, Hu R F, Sun Q H. Influence of rigidity of feed system with ball screw in NC lathe on positioning precision [J]. Engineering Science, 2004, 6 (9): 46-49.

[9] Esmailzaden E, Ohadt A R. Vibration and stability analysis of non-uniform Timoshenko beams under axial and distributed tangential loads [J]. Journal of Sound and Vibration, 2000, 236 (3): 443-456.

[10] Arboleda M I G, Zapata M D G, Aristizabal O J D. Stability and natural frequencies of weakened Timoshenko bean-column with generalized end conditions under constant axial load [J]. Journal of Sound and Vibration, 2007 (307): 89-112.

[11] Wang L H, Du R S, Wu B, et al. Non-linear dynamic characteristics of NC table [J]. China Mechanical Engineering, 2009, 20 (13): 1513-1519.

[12] Wang L H, Du R S, Wu B, et al. Chaotic characteristic analysis on dynamic properties of NC table [J]. China Mechanical Engineering, 2009, 20 (14): 1656-1659.

# Chapter 8 Adaptive Control of Feed System Based on LuGre Model

## 8.1 Introduction

Because of the existence of disturbance such as non-linear friction, ball screw pair in feed system can not fill the high precision bill with the traditional control method[1-2]. Compensation based on friction model is a widely used method. The key of this method is to establish an accurate friction model[3-4]. The value of the system friction is calculated according to the velocity, position, and so on. Then the effect of the friction can be offset by inputting an equivalent output moment into control moment. The friction model can be divided into static friction model and dynamic friction model. At present, the LuGre model is widely used. The LuGre model describes the dynamic and static characteristics accurately and it has a perfect compensation effect. Kamalzadeh and Erkorkmaz[5] proposed an adaptive sliding mode controller for the vibration characteristics at the axial of the ball screw pair feed system. But this controller does not contain the model which can reflect the friction characteristics of the system accurately. Choi[6] established a dynamic friction model to represent the friction hysteresis according to the LuGre model. This dynamic friction model is used in the system sliding mode controller and the system has higher tracking characteristics. However, the Preisach model and the neural network algorithm make this control method more complex. Zhou et al.[7] realized the friction compensation and the load disturbance estimation and just used one coefficient to reflect the parameters' change of the LuGre model, but it is not accurate. Yang et al.[8] proposed a method which can realize repeating adaptive friction compensation. This method can improve the position tracking accuracy of the feed system and uses two non-linear observers to estimate the value of parameter $z$. If the accuracy is enough, the method will have an effect on response speed. In fact, the friction compensation for the feed system has become an important issue and one of the research hotspots in the feed system vibration controlling.

## 8.2 Dynamic modeling of the feed system

The mechanical part of the system can be assumed to be the worktable part which moves in a straight line and the motor shaft-screw part which moves in rotating motion. Then the kinematical equation of the worktable can be obtained:

$$m\ddot{x} = F_D - F_f - F_r \qquad (8-1)$$

where, $m$ is the weight of the worktable; $x$ is the axial displacement of the worktable; $F_D$ is the driving force applied to the worktable; $F_f$ is the friction applied to the worktable; and $F_r$ is the external disturbance force applied to the worktable, including the cutting force and so on.

The kinematical equation of the motor shaft-screw part can be obtained:

$$J_m \ddot{\theta}_m = Ku - D\dot{\theta}_m - F_d R \qquad (8-2)$$

where, $J_m$ is the equivalent rotational inertia of feed system; $\theta_m$ is the rotation angle of the motor shaft-screw; $K$ is the motor torque constant; $u$ is the controlled variable; $D$ is the equivalent damping coefficient of the feed system; $F_d$ is the load force applied to the screw; and $R$ is the radius of the screw.

The transformational relation between the $F_D$ in equation (8−1) and the $F_d$ in equation (8−2) can be described as

$$\frac{F_D}{F_d} = \frac{2\pi R}{h} \qquad (8-3)$$

where, $h$ is the lead of the screw.

The dynamic equation of the system can be obtained according to equation (8−1) ~ equation (8−3):

$$J\ddot{\theta}_m = Ku - D\dot{\theta}_m - M_f - T_r \qquad (8-4)$$

where, $J = J_m + \dfrac{h^2}{4\pi^2} \cdot m$; $M_f = \dfrac{h}{2\pi} \cdot F_f$; $T_r = \dfrac{h}{2\pi} \cdot F_r$.

The LuGre dynamic model[9] can describe every dynamic and static characteristic of the friction accurately, such as sick-slip, limit-cycle oscillation, former sliding deformation, friction memory, changed static friction and static Strobeck curve.

According to the friction compensation which is based on the LuGre friction model, the friction moment applied to the system can be described as

$$\begin{cases} \dot{z} = \dot{\theta} - \dfrac{|\dot{\theta}|}{g(\dot{\theta})} z \\ \sigma_0 g(\dot{\theta}) = M_C + (M_s - M_C)\exp(-\dot{\theta}_s^2) \\ M_f = \sigma_0 z + \sigma_1 \dot{z} + \mu_b \dot{\theta} \end{cases} \qquad (8-5)$$

where, $\dot{z}$ is the average deformation dynamic equation of the mane surface; $\dot{\theta}$ is the relative velocity of the contact surface; and $z$ is the average deformation which can not be measured directly in fact. When non-linear function $g(\dot{\theta}) > 0$, the different frictional effect is reffected, where $M_C$ is the Coulomb friction moment, $M_s$ is the maxim static friction moment, and $\dot{\theta}_s$ is Stribeck velocity. In the $M_f$, $\sigma_0$ is the stiffness coefficient of the mane, $\sigma_1$ is the damping coefficient of the mane, and $\mu_b$ is the viscidity frictional coefficient.

During the actual application, the fiction moment can be changed by the LuGre model's parameters which are influenced by the temperature, lubrication and material wear. Then the LuGre fiction model will be divided into two parts to make the control system obtain a wonderful adaptive ability. One of the two parts is about parameter $z$ whose changing will be reflected by using $\alpha$. Another one is about the viscidity frictional coefficient whose changing will be reflected by using $\beta$. So the friction moment $M_f$ in the LuGre model can be modified as

$$M_f = \alpha(\sigma_0 z + \sigma_1 \dot{z}) + \beta \mu_b \dot{\theta} \qquad \alpha > 0, \beta > 0 \qquad (8-6)$$

Combining equation (8-4) and equation (8-6) and considering $\dot{\theta} = k_0 \dot{\theta}_m$, where $k_0$ is the transformation coefficient between load revolving speed $\dot{\theta}$ and motor revolving speed $\dot{\theta}_m$, the relationship can be obtained:

$$J\ddot{\theta}_m = Ku - D\dot{\theta}_m - T_r - \alpha(\sigma_0 z + \sigma_1 \dot{z}) - k_0 \beta \mu_b \dot{\theta}_m \qquad (8-7)$$

## 8.3 Designing of the friction compensation controller

Due to the poor robustness, ordinary PID controllers cannot meet the high accuracy requirements[10]. Backstepping method has a unique superiority in dealing with non-linear control problems. This method cancels the constraints of the matching conditions which are satisfied with the indeterminacy of the system, so the method solves the problem of controlling in the relatively complicated non-linear system. Therefore, based on these features, the Backstepping method will be used to design the adaptive friction compensation controller.

A non-linear observation will be used in a simplified way to observe the value of $z$ in the LuGre model and the equation is as follows:

$$\dot{\hat{z}} = \dot{\theta} - \frac{|\dot{\theta}|}{g(\dot{\theta})}\hat{z} + k_z e_2 \left[\sigma_0 - \sigma_1 \frac{|\dot{\theta}|}{g(\dot{\theta})}\right] \quad (8-8)$$

where, $k_z$ is a positive real number.

Assuming $\tilde{z}$ is the error of the state observer, it can be obtained as follows:

$$\tilde{z} = z - \hat{z} \quad (8-9)$$

$$\dot{\tilde{z}} = -k_z e_2 \left[\sigma_0 - \sigma_1 \frac{|k_0 \dot{\theta}_m|}{g(k_0 \dot{\theta}_m)}\right] - \frac{|k_0 \dot{\theta}_m|}{g(k_0 \dot{\theta}_m)}\tilde{z} \quad (8-10)$$

Defining $e_1$ as the error of the position tracking in the feed system, it can be obtained as follows:

$$e_1 = \theta_e - \theta_m \quad (8-11)$$

where, $\theta_e$ is the expectation angel displacement.

Then the dynamic equation of error $\dot{e}_1$ is as follows:

$$\dot{e}_1 = \dot{\theta}_e - \dot{\theta}_m \quad (8-12)$$

There is a need to solve the state equation of the system by using the Lyapunov function. Generally, the stability of a system is estimated by creating a scalar function and directly studying the symbolic characteristics of its derivative. It is easy to output the error $e_1$ and obtain the expected value of $\dot{\theta}_m$. And the $V_1$ will be chosen as follows:

$$V_1 = \frac{1}{2}e_1^2 \quad (8-13)$$

$$\dot{V}_1 = e_1 \dot{e}_1 = e_1 (\dot{\theta}_e - \dot{\theta}_m) \qquad (8-14)$$

Reference velocity signal can be designed by using Backstepping design idea as follows:

$$\begin{cases} \theta_{1d} = \dot{\theta}_e + k_1 e_1 + k_2 \chi \\ \chi = \int_0^t e_1 \mathrm{d}t \end{cases} \qquad (8-15)$$

where, $\theta_{1d}$ is the control signal of velocity; $k_1 > 0$; $k_2 > 0$; $\chi$ is the integration of position tracking error, it can ensure that the system tracking error can approach zero when the load or model is indeterminate.

Because there is an error $e_2$ between the controlled variable of reference velocity $\theta_{1d}$ and the actual velocity $\dot{\theta}_m$, error $e_2$ can be described as follows:

$$e_2 = \theta_{1d} - \dot{\theta}_m = \dot{\theta}_e - \dot{\theta}_m + k_1 e_1 + k_2 \chi \qquad (8-16)$$
$$\dot{e}_2 = \ddot{\theta}_e - \ddot{\theta}_m + k_1 \dot{e}_1 + k_2 e_1 \qquad (8-17)$$

Combining equation $(8-12)$ and equation $(8-16)$, the following equation can be obtained:

$$\dot{e}_1 = e_2 - k_1 e_1 - k_2 \chi \qquad (8-18)$$

Then the system dynamic equation can be described as follows:

$$J \dot{e}_2 = J(\ddot{\theta}_e + k_1 \dot{e}_1 + k_2 e_1) - Ku + D\dot{\theta}_m + T_r + \alpha z \left[ \sigma_0 - \sigma_1 \frac{|k_0 \dot{\theta}_m|}{g(k_0 \dot{\theta}_m)} \right] +$$
$$\alpha \sigma_1 k_0 \dot{\theta}_m + \beta \mu_b k_0 \dot{\theta}_m \qquad (8-19)$$

Because $\alpha$ and $\beta$ are not certain parameters in the system, there is a need to estimate the value of $\alpha$ and $\beta$ as $\bar{\alpha}$, $\bar{\beta}$ by using the adaptive method. Then the error of $\alpha$, $\beta$ can be described as follows:

$$\begin{cases} \tilde{\alpha} = \alpha - \bar{\alpha} \\ \tilde{\beta} = \beta - \bar{\beta} \end{cases} \qquad (8-20)$$

The control law can be designed for the system as follows:

$$Ku = \bar{J}(\ddot{\theta}_e + k_1 \dot{e}_1 + k_2 e_1) + \frac{e_1}{k_3} + k_c e_2 + \bar{D}\dot{\theta}_m + \bar{T}_r + \bar{\alpha} \bar{z} \left[ \sigma_0 - \sigma_1 \frac{|k_0 \dot{\theta}_m|}{g(k_0 \dot{\theta}_m)} \right] +$$
$$\bar{\alpha} \sigma_1 k_0 \dot{\theta}_m + \bar{\beta} \mu_b k_0 \dot{\theta}_m \qquad (8-21)$$

where, $k_3$ is a positive real number; $\bar{J}$ is the observation value of parameter $J$; $\bar{D}$ is the observation value of parameter $D$.

Then Lyapunov function can be defined as follows:

$$V_2 = V_1 + \frac{1}{2} k_2 \chi^2 + \frac{1}{2} k_3 J e_2^{\ 2} + \frac{1}{2\gamma_1} \tilde{\alpha}^2 + \frac{1}{2\gamma_2} \tilde{\beta}^2 + \frac{1}{2\gamma_3} \tilde{T}_r^{\ 2} + \frac{1}{2\gamma_4} \tilde{D}^2 + \frac{1}{2} \alpha \tilde{z}^2$$
$$(8-22)$$

$$\dot{V}_2 = e_1\dot{e}_1 + k_2\chi e_1 + k_z e_2 J\dot{e}_2 + \frac{1}{\gamma_1}\tilde{\alpha}\dot{\tilde{\alpha}} + \frac{1}{\gamma_2}\tilde{\beta}\dot{\tilde{\beta}} + \frac{1}{\gamma_3}\tilde{T}_r\dot{\tilde{T}}_r + \frac{1}{\gamma_4}\tilde{D}\dot{\tilde{D}} + \alpha z\dot{\tilde{z}} =$$

$$k_z e_2 \left\{ J\ddot{\theta}_e + Jk_1\dot{e}_1 + Jk_2 e_1 - Ku + D\dot{\theta}_m + \alpha z\left[\sigma_0 - \sigma_1\frac{|k_0\dot{\theta}_m|}{g(k_0\dot{\theta}_m)}\right] + \alpha\sigma_1 k_0\dot{\theta}_m + \right.$$

$$\beta\mu_b k_0\dot{\theta}_m + T_r \Big\} + k_2\chi e_1 + e_1(e_2 - k_1 e_1 - k_2\chi) + \frac{1}{\gamma_1}\tilde{\alpha}\dot{\tilde{\alpha}} + \frac{1}{\gamma_2}\tilde{\beta}\dot{\tilde{\beta}} + \frac{1}{\gamma_3}\tilde{T}_r\dot{\tilde{T}}_r +$$

$$\frac{1}{\gamma_4}\tilde{D}\dot{\tilde{D}} - \alpha\tilde{z}\left\{\frac{|k_0\dot{\theta}_m|}{g(k_0\dot{\theta}_m)}\tilde{z} + k_z e_2\left[\sigma_0 - \sigma_1\frac{|k_0\dot{\theta}_m|}{g(k_0\dot{\theta}_m)}\right]\right\} \qquad (8-23)$$

where, $\gamma_1$, $\gamma_2$ and $\gamma_3$ are the adaptive gain; $\tilde{T}_r = T_r - \overline{T}_r$; $\tilde{D} = D - \overline{D}$.

Combining equation (8−21) and equation (8−23) and considering $\alpha z - \overline{\alpha}\,\tilde{z} = \alpha\tilde{z} + \tilde{\alpha}\,\tilde{z}^{[11-12]}$, the following equation can be obtained:

$$\dot{V}_2 = -k_1 e_1^2 - k_z k_c e_2^2 - \alpha\frac{|k_0\dot{\theta}_m|}{g(k_0\dot{\theta}_m)}\tilde{z}^2 + \tilde{T}_r\left(k_z e_2 - \frac{1}{\gamma_3}\dot{\tilde{T}}_r\right) + \tilde{D}\left(k_z e_2\dot{\theta}_m - \frac{1}{\gamma_4}\dot{\tilde{D}}\right) +$$

$$\tilde{\beta}\left(k_z e_2\mu_b k_0\dot{\theta}_m - \frac{1}{\gamma_2}\dot{\tilde{\beta}}\right) + \tilde{\alpha}\left\{k_z e_2\tilde{z}\left[\sigma_0 - \sigma_1\frac{|k_0\dot{\theta}_m|}{g(k_0\dot{\theta}_m)}\right] + k_z e_2\sigma_1 k_0\dot{\theta}_m - \frac{1}{\gamma_1}\dot{\tilde{\alpha}}\right\}$$

$$(8-24)$$

Chose the following adaptive law:

$$\begin{cases} \dot{\tilde{\alpha}} = \gamma_1 k_z e_2\left[\tilde{z}\sigma_0 - \tilde{z}\sigma_1\frac{|k_0\dot{\theta}_m|}{g(k_0\dot{\theta}_m)} + \sigma_1 k_0\dot{\theta}_m\right] \\ \dot{\tilde{\beta}} = \gamma_2 k_z e_2\mu_b k_0\dot{\theta}_m \\ \dot{\tilde{T}}_r = \gamma_3 k_z e_2 \\ \dot{\tilde{D}} = \gamma_4 k_z e_2\dot{\theta}_m \end{cases} \qquad (8-25)$$

Combing equation (8−24) and equation (8−25), the following equation can be obtained:

$$\dot{V}_2 = -k_1 e_1^2 - k_z k_c e_2^2 - \alpha\frac{|k_0\dot{\theta}_m|}{g(k_0\dot{\theta}_m)}\tilde{z}^2 \leqslant 0 \qquad (8-26)$$

In conclusion, according to the Lyapunov theory, the system can be ensured to become globally asymptotically stable by using adaptive friction compensation control law with equation (8−21) and equation (8−25).

## 8.4 Simulating of adaptive friction compensation

MATLAB/SIMULINK will be used to verify the validity of the integration back-stepping adaptive compensation arithmetic[13—15]. The simulating diagram is shown in Fig. 8—1 and Fig. 8—2.

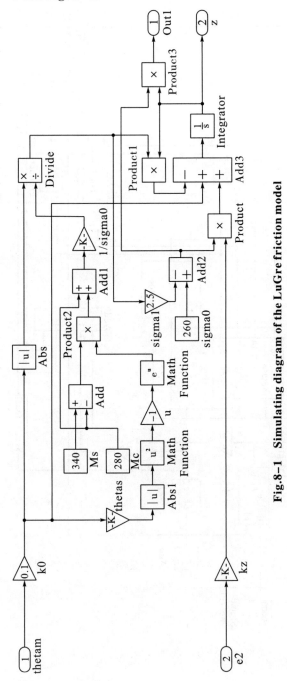

Fig.8—1 Simulating diagram of the LuGre friction model

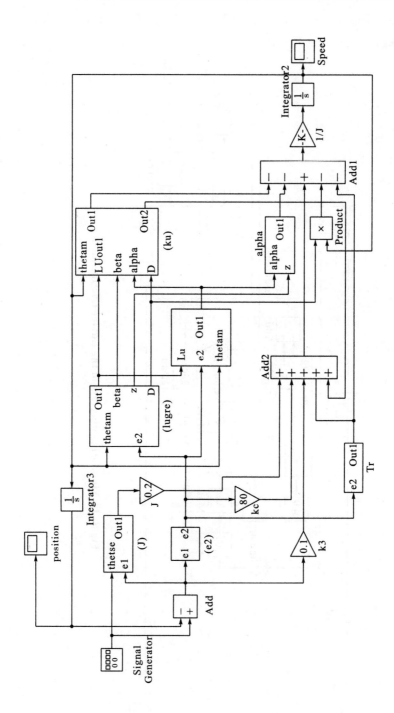

Fig.8-2  Simulating diagram of the adaptive friction compensation system

## Chapter 8 Adaptive Control of Feed System Based on LuGre Model

The system has been simulated in this book to verify the validity of the adaptive friction compensation controlling. The simulating parameters of LuGre model were chosen as follows: $M_s=340$ N, $M_C=280$ N, $\mu_b=0.02$ N·s/m, $\sigma_0=260$ N/m, $\sigma_1=2.5$ N·s/m, $\theta_S=0.01$ m/s, $J=0.2$ kg·m². In the simulating, the adaptive parameters were changed frequently to get better tracking results. The initial value of the adaptive parameters was chosen as follows: $k_c=80$, $k_1=13$, $k_2=1$, $k_3=10$, $k_0=0.1$, $k_z=0.1$, $\gamma_1=2$, $\gamma_2=10$, $\gamma_3=10$, $\gamma_4=5$. The system was simulated with these parameters and the input signal was the sinusoidal signal.

The system was simulated to verify the validity of the initial values of the adaptive coefficients and the simulating time was set to 20 s. The velocity tracking and position tracking results are shown in Fig. 8−3.

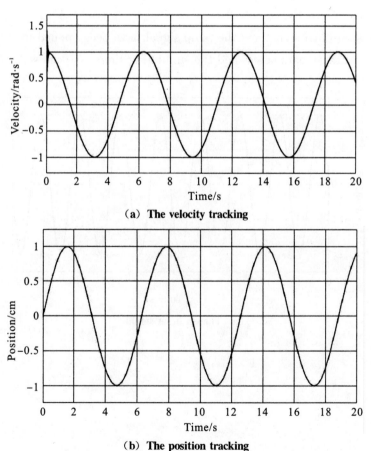

(a) The velocity tracking

(b) The position tracking

Fig. 8−3 The velocity and position tracking result

In Fig. 8−3(a), there is chattering in the velocity tracking diagram at the start moment. To restrain this chattering phenomenon, adaptive coefficient parameters were changed as follows: $k_c=80$, $k_1=10$, $k_2=1$, $k_3=13$, $k_0=0.1$, $k_z=0.102$, $\gamma_1=2$, $\gamma_2=100$, $\gamma_3=100$, $\gamma_4=75$. The velocity tracking result of the system simulating is shown in Fig. 8−3(b). The diagram shows that the chattering is restrained at the start moment and the tracking result gets better when the value of $k_z$ is between 0.1 and 0.12, and the value of $\gamma_1$ is between 1.9 ~ 2.1. In this simulating, $k_z$ is the error compensation coefficient of the observation and this parameter will have an indirect effect on the accuracy of the LuGre model and the compensation of the friction moment; $\gamma_1$ is the variation coefficient of the non-linear part of LuGre model and this parameter determines the accuracy of the friction which is estimated by the friction model directly.

The sinusoidal signal and the ramp signal have been applied to the system according to these parameters and the simulating time has been set as 50 s.

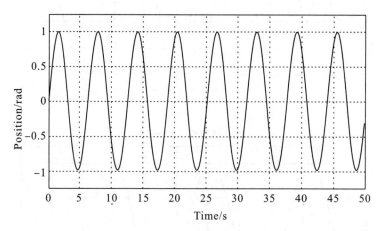

(a) **The sin input signal**

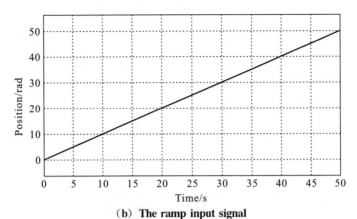

(b) The ramp input signal

Fig. 8−4  The input signal of the system

The friction moment tracking curve and system position error tracking curve are shown in Fig. 8−5 and Fig. 8−6 when different signals are input into the system. In Fig. 8−5, the system position error decreases from $-7.2\times 10^{-3}$ rad to $-0.25\times 10^{-3}$ rad in 0.3 s and finally it stabilizes between $-0.17\times 10^{-4}$ rad and $0.17\times 10^{-4}$ rad. In Fig. 8−6, the system position error decreases from $-7.2\times 10^{-3}$ rad to $-0.4\times 10^{-3}$ rad in 0.3 s. During this whole process, there is no higher chattering and the output velocity becomes stable after 0.24 s. Therefore a higher tracking accuracy and faster response speed can be obtained. The friction moment changes along with the variation of the speed and finally gets stable[16]. At the same time, the system obtains better robustness when there is an outside disturbance.

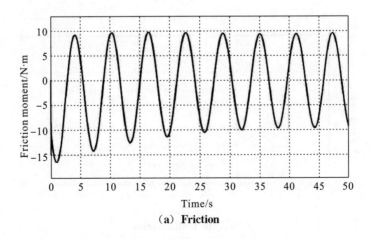

(a) Friction

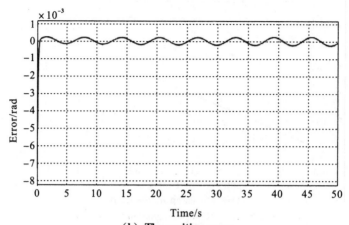

(b) The position error

Fig. 8-5　Sinusoidal signal

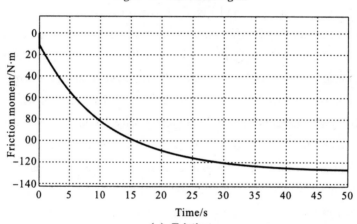

(a) Friction

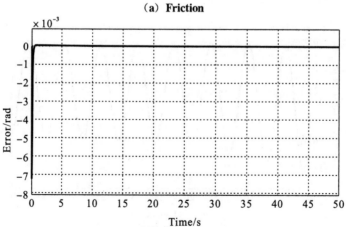

(b) The position error

Fig. 8-6　Ramp signal

# References

[1] Wang L, Liu H T, Liang T, et al. Modeling and analysis of dynamic response of servo feed system under low frequency excitation [J]. Chinese Journal of Mechanical Engineering, 2015, 51 (3): 80-86.

[2] Huang L L, Zha C L. Modeling and analysis of NC machine tool transmission system [J]. Manufacturing Automation, 2015, 37 (7): 92-95.

[3] Gordon D J, Erkorkmaz K. Accurate control of ballscrew drives using pole-placement vibration damping and a novel trajectory prefilter [J]. Precision Engineering, 2013, 37 (2): 308-322.

[4] Brain A H, Pieere D, Carlos C D W. A survey of models, analysis tools and compensation methods for the control of machines with friction [J]. Automatica, 1994, 30 (7): 1083-1138.

[5] Kamalzadeh A, Erkorkmaz K. Compensation of axial vibrations in ball screw drives [J]. Annals of the CIRP, 2007, 56 (1): 373-378.

[6] Choi J. Development of hysteresis friction model for a precise tracking control system [J]. Proceedings of the Institution of Mechanical Engineers Part C — Journal of Mechanical Engineering Science, 2012, 226 (5): 1338-1344.

[7] Zhou J Z, Duan B Y, Huang J. Effect and compensation for servo systems using LuGre friction model [J]. Control Theory & Applications, 2008, 25 (6): 990-994.

[8] Yang S, Zeng M, Su B K. New repetitive adaptive friction compensation scheme in high-precise servo system [J]. Journal of Southeast University (Natural Science Edition), 2006, 36 (1): 74-78.

[9] Ashwani K P, Jin H O, Dennis S B. On the LuGre model and friction induced hysteresis [C]. Minneapolis: Proceedings of the 2006 American Control Conference, 2006 (8): 3247-3252.

[10] Liu D S, Bai H, Li W J, et al. Analysis and simulation for self-adaptive control method based on LuGre friction model [J]. Computing Technology and Automation, 2015, 34 (3): 16-20.

[11] Zhang D. Compensation control of servo system with friction [D]. Xi'an: Xidian University, 2008.

[12] Do N B, Ferri A A, Bauchau O A. Efficient simulation of a dynamic system with LuGre friction [J]. Journal of Computational and Non-linear Dynamics, 2007, 2 (4): 281-289.

[13] Freidovich L, Robertsson A, Shiriaev A, et al. LuGre-model-based friction

compensation [J]. IEEE Transactions on Control Systems Technology, 2010, 18 (1): 194-200.

[14] Barabanov N, Ortega R. Necessary and sufficient conditions for passivity of the LuGre friction model [J]. IEEE Transactions on Automatic Control, 2000, 45 (4): 830-832.

[15] Wang X D, Jiao Z X, Xie S C. LuGre-based compensation for friction in electro-hydraulic loading [J]. Journal of Beijing University of Aeronautics and Astronautics, 2008, 34 (11): 1254-1257.

[16] Liu C C, Tsai M S, Cheng C C. Development of a novel transmission engaging model for characterizing the friction behavior of a feed drive system [J]. Mechanism and Machine Theory, 2019 (134): 425-439.